T0092102

Digital Lifeline?

Information Policy Series

Edited by Sandra Braman

The Information Policy Series publishes research on and analysis of significant problems in the field of information policy, including decisions and practices that enable or constrain information, communication, and culture irrespective of the legal siloes in which they have traditionally been located, as well as state-law-society interactions. Defining information policy as all laws, regulations, and decision-making principles that affect any form of information creation, processing, flows, and use, the series includes attention to the formal decisions, decision-making processes, and entities of government; the formal and informal decisions, decision-making processes, and entities of private and public sector agents capable of constitutive effects on the nature of society; and the cultural habits and predispositions of governmentality that support and sustain government and governance. The parametric functions of information policy at the boundaries of social, informational, and technological systems are of global importance because they provide the context for all communications, interactions, and social processes.

Virtual Economies: Design and Analysis, Vili Lehdonvirta and Edward Castronova

Traversing Digital Babel: Information, e-Government, and Exchange, Alon Peled

Chasing the Tape: Information Law and Policy in Capital Markets, Onnig H. Dombalagian

Regulating the Cloud: Policy for Computing Infrastructure, edited by Christopher S. Yoo and Jean-François Blanchette

Privacy on the Ground: Driving Corporate Behavior in the United States and Europe, Kenneth A. Bamberger and Deirdre K. Mulligan

How Not to Network a Nation: The Uneasy History of the Soviet Internet, Benjamin Peters

Hate Spin: The Manufacture of Religious Offense and its Threat to Democracy, Cherian George

Big Data Is Not a Monolith, edited by Cassidy R. Sugimoto, Hamid R. Ekbia, and Michael Mattioli

Decoding the Social World: Data Science and the Unintended Consequences of Communication, Sandra González-Bailón

Open Space: The Global Effort for Open Access to Environmental Satellite Data, Mariel John Borowitz

Digital Lifeline?

ICTs for Refugees and Displaced Persons

Edited by Carleen F. Maitland

The MIT Press
Cambridge, Massachusetts
London, England

© 2018 Massachusetts Institute of Technology

All rights reserved. No part of this book may be reproduced in any form by any electronic or mechanical means (including photocopying, recording, or information storage and retrieval) without permission in writing from the publisher.

This book was set in Stone Serif by Westchester Publishing Services.

Library of Congress Cataloging-in-Publication Data

Names: Maitland, Carleen F., editor.
Title: Digital lifeline? : ICTs for refugees and displaced persons / edited by Carleen F. Maitland.
Description: Cambridge, MA : MIT Press, [2018] | Series: Information policy series | Includes bibliographical references and index.
Identifiers: LCCN 2017038912 | ISBN 9780262535083 (pbk. : alk. paper)
Subjects: LCSH: Refugees--Information services. | Information technology--Social aspects.
Classification: LCC HV640 .D54 2018 | DDC 362.87/8--dc23
LC record available at https://lccn.loc.gov/2017038912

Contents

Series Editor's Introduction

Sandra Braman

The U.S. Federal Communications Commission (FCC) has a long history of allocating spectrum at the very margins of usability to ham radio operators. "Hammers," amateur operators, use the electromagnetic spectrum that carries broadcast radio and television, cellphone signals, and other types of wireless transmission to communicate non-commercially across long distances, often experimenting in ways that push the technical boundaries of the medium. The regulator learned long ago that these hobbyists regularly produce innovations that then allow the same portions of the spectrum to be used for commercial purposes; the reallocation process then moves ham radio operators further out into what remains unusable, for the cycle to repeat again.

While ham radio operators work on the technological margins of communicability, others find themselves on the social margins. The problem of inequities in access to technologies within and across societies, often referred to as the digital divide, has long been acknowledged, but for refugees and displaced persons the situation is particularly extreme. As of 2016, according to the office of the United Nations High Commissioner for Refugees, there were almost 90 million refugees (who must move internationally) and displaced persons (forced to move within a country), as well as another 10 million people who were stateless, with the numbers growing significantly every day. The causes are multiple: war and civil strife; religious and racial hatred; famine, drought, rising waters, and other consequences of climate change; failed economies and social structures. The humanitarian needs are stupendous, heartbreaking, some of the most important moral, political, economic, and logistical challenges of our time. Among them are the ability to communicate and access to information, often necessities before other needs can be addressed.

This edited collection on efforts by organizations working with refugees and displaced persons to address their communicative and informational needs provides a foundation for those who want to understand the current status of those efforts and those who would do research on these developments. The book approaches the problem from several directions, both social and technical, with running attention to the information policy issues these undertakings present throughout. It provides such basics as how the status of refugees is determined, what the legal framework for working with these populations is, the extent to which networks penetrate the environments in which refugees and displaced persons are sheltered, the roles of particular types of technologies and networks, the kinds of information needed by refugees and displaced persons, and the challenges those organizations face in working with these populations. The communication and information challenges are so great that, as the book's editor Carleen Maitland points out, the concept of the "digital refugee" has come into use.

Wide ranging as it is, what this book offers is just the beginning of the kind of research, thinking, experimentation, policy-making, and practice needed. What are the downsides of the technologies provided? The same systems can serve surveillance purposes of concern to everyone from a privacy perspective and of enormous immediate concern to many refugees and displaced persons for political reasons—so much so that these may well be called "identity technologies" rather than "information and communication technologies" (ICTs). What are users actually doing with these technologies, and how do they experience official uses? How are those in camps and on the move adapting the technologies to their own ends? What is the political economy of the provision of, for example, cell phones to refugees and displaced persons—who wins, economically, and who loses? At what rate are communication networks put in place to serve these populations growing relative to those of society at large? How is decision-making for all of this taking place? And how do the practices of those trying to welcome refugees differ from those trying to send them away? The U.S. practice of taking cell phones away from individuals being deported to Mexico (along with their other goods) provides a depressing counterpart to the efforts reported upon here.

As the effects of climate change become more evident and extreme, and as the political inversion driving so many societies to the far right or to chaos altogether continues, it may be that research on the provision and use of information and communication technologies to those being handled by the international humanitarian system will provide insights into a future for all of us. A final research question needing attention by those studying the uses of information and communication technologies by those institutions working with refugees and displaced persons would be what this work is yielding in technology and network design, institutional management of communications and information, and treatment of information policy issues in liminal spaces. Will the technological, use, and institutional innovations being developed to deal with these socially marginal populations ultimately be used for commercial (and political) purposes that affect us all, as has happened so often with the innovations of technically marginal ham radio operators?

Illustrations

Acknowledgments

This project is the result of the efforts, support, and patience of a number of people over the past five years. Starting with our first workshop, in Stellenbosch, South Africa in 2013, I want to thank those participants who worked tirelessly during two days of deliberations, grappling with the technological advances and humanitarian crises addressed in these pages. I am particularly appreciative of the practitioners who attended, Chad Wesen of the U.S. State Department, Kathryn Hoeflich of the Cape Town Refugee Centre, and Rosa Akbari of DiMagi, who in a spirit of sharing, provided immense benefit and insights to the academic participants. We were saddened by the subsequent loss of participant Dr. Gaetano Borriello, whose contributions during the workshop and over the course of his career have benefitted the displaced worldwide.

Contributions to this volume also derived from our second workshop, held in Washington, DC in 2016, which would not have been possible without the support of John Pope of the PeaceTech Lab and Jeff Landale of the X-lab at Penn State University. Our discussions also benefitted from the insights of Scott Edwards of Amnesty International. Both workshops were made possible by financial contributions of the U.S. National Science Foundation (award #1343520) and Penn State University's Institute for Information Policy.

The book also draws upon fieldwork facilitated by, or done in partnership with, a number of amazing people from the humanitarian community. From Za'atari Camp in Jordan, I am more than deeply indebted to UNHCR's Irene Omondi and her staff, whose passion, commitment, and patience seems to know no bounds. I am also thankful to former UNHCR camp director Hovig Etyemezian, who together with Irene, embraced innovation and engagement with academics. I am also very thankful for the Za'atari staff of IRD, for their hard work and gracious hospitality. Also from UNHCR, the regional protection as well as IT staff have been very helpful, including but not limited to Jean Laurent Martin, Edouard Legoupil, and the 'invisible' Manal Stulgaitis, with whom I've had the pleasure of working both in Cape Town and Amman, although we have never met in person. I also want to extend thanks to new UNHCR collaborators in Rwanda, Ruslan Shabdunov and Amal al Beedh, who in just a short period of time have shown themselves to be intellectually curious, energetic collaborators. Finally, and most importantly, I want to thank all of the refugees who took the time to share their thoughts, opinions, and stories about their use of ICTs, demonstrating the many experiences we share.

I also owe thanks to my academic colleagues. Many of the insights reflected in these chapters would not have been possible without the assistance, insight, humor, and hospitality of Drs. Nijad al-Najdawi of Al Balqa University and Sara Tedmori of Princess Sumaya University of Technology in Jordan. Dr. al-Najdawi, a participant in our Stellenbosch workshop, was instrumental in helping us gain access to Za'atari camp. Together, he and Sara subsequently became valued collaborators and opened their home on many occasions, introducing us to Jordanian life. Our work in Jordan also benefitted from National Science Foundation support (award # 1427873). More recently, in the spring of 2017, I attended two excellent workshops on ICTs and migration, hosted, respectively, by Dr. Arul Chib at Nanyang Technological University and Dr. Saskia Witteborn, both from the Chinese University of Hong Kong. The workshops were very helpful in crystallizing ideas as well as gaining insights into the experiences of asylum-seekers in Asia.

This work would not have been possible without my wonderful and dedicated team of doctoral students, Eric Obeysekare, Dan Hellmann, Rich Caneba, and Ying Xu, as well as undergraduate honors student Katelyn Sullivan. Whether directly or indirectly, they have all contributed in large and small ways to this endeavor. Rich has been extremely helpful in preparation of the manuscript, Eric has provided valuable assistance in Rwanda, and Ying has served as 'co-conspirator' on much of the field research. I also want to thank my College and University for their support both financial and otherwise for these endeavors.

The staff at MIT Press and three anonymous reviewers have contributed immensely to this project by pushing me and my fellow authors to give voice to our deep engagement in this field. The manuscript was significantly improved due to their careful reviews. I also owe many thanks to Information Policy series editor Sandra Braman, who has been incredibly supportive of this volume's innovative and somewhat unique approach.

Finally, and most importantly, I owe a huge debt of gratitude to my family. My parents and sisters Kelly Sikorski, Michelle Maitland, and Mary Beth Pettigrew have always supported me in ways that makes it possible to do this sometimes difficult work. My daughter Zoe has been incredibly patient and flexible through the travel and deep thought this project required. My deepest debt of gratitude goes to my partner Steve Mower. He has been my rock, my supporter, my fan, and, perhaps, most importantly, my strongest, yet kindest, critic. His careful reading and suggestions for edits, and all he does in our daily lives, particularly his humor, are invaluable.

Abbreviations

ACLED	Armed Conflict Location and Event Dataset
ACM CHI	Association for Computing Machinery Conference on Human Factors in Computing Systems
AFC	Alternate Food Collectors
API	Application Programming Interface
ARC	UK Home Office's Applicant Registration Card
ASA	Authorized Shared Access
ATM	Automated Teller Machine
BIMS	Biometric Identity Management System
BTS	Base Station
CARFMS	Canadian Association for Refugee and Forced Migration Studies
CERF	Central Emergency Response Fund
CIO	Chief Information Officer
COI	Country of Origin Information
COMEST	World Commission on the Ethics of Scientific Knowledge and Technology
CRM	Customer Relationship Management
DFID	Department for International Development
DfNR	Directorate for Nationality and Refugees
DHS	Department of Homeland Security
DoJ	Department of Justice
DQA	Data Quality Assessment
DRA	Department of Refugee Affairs
DSA	Dynamic Spectrum Access
EDE	Electronic Data Exchange

EMP UP	Empathy Up
EO	Earth Observation
EOS	Expandable Open Source
EU	European Union
FDP	Final Delivery Point
FEWS	Famine Early Warning Systems
FSA	Free Syrian Army
GAA	General Authorized Access
GBV	Gender-Based Violence
GBVIMS	Gender-Based Violence Information Management System
GeoIMS	Geographic Information Management System
GDP	Gross Domestic Product
GDT	Global Distribution Tool
GIS	Geographic Information System
GPS	Global Positioning System
GSM	Global System for Mobile Communication
HCI	Human-Computer Interaction
HCI4D	Human-Computer Interaction for Development
HPPP	UK Border Agency's Human Provenance Pilot Project
HSPA	High-Speed Packet Access
IASFM	International Association for the Study of Forced Migration
ICT	Information and Communication Technology
IDP	Internally Displaced Person
IGO	Intergovernmental Organization
IMS	Internet Protocol Multimedia Subsystem
INGO	International Non-Governmental Organization
IOM	International Organization for Migration
IP	Internet Protocol
IRC	International Rescue Committee
IRD	International Relief and Development
ISIL	Islamic State of Iraq and the Levant
ISIS	Islamic State of Iraq and Syria
ISM	Industrial, Scientific, and Medical
ISP	Internet Service Provider
IT	Information Technology
ITU	International Telecommunications Union
JRS	Jesuit Refugee Service

LAN	Local Area Network
LGBT	Lesbian, Gay, Bisexual, and Transgender
LSA	Licensed Shared Access
LTE	Long-Term Evolution
M&E	Monitoring and Evaluation
MDI	Mass Data Institute (at Georgetown University)
MEPS	Middle East Payment Systems
MOU	Memorandum of Understanding
MS	Mobile Station
NGO	Non-Governmental Organization
NIE	New Institutional Economics
NOAA	National Oceanographic and Atmospheric Administration
NSDI	National Spatial Data Infrastructure
NSF	National Science Foundation
OCHA	Office of Coordination in Humanitarian Affairs
ODK	Open Data Kit
OGP	Open Government Partnership
OIG	Office of Inspector General
PA	Priority Access
PBX	Private Branch Exchange
PC	Personal Computer
PII	Personally Identifiable Information
PMP	Performance Management Plan
PPR	Performance and Planning Reports
PRM	Bureau of Population, Migration, and Refugees
PSTN	Public Switched Telephone Network
Q&A	Question and Answer
QoS	Quality of Service
RAIS	Refugee Assistance Information System
RF	Radio Frequency
RIT	Rochester Institute of Technology
RSD	Refugee Status Determination
RTP	Real-Time Transport Protocol
SDCCH	Stand-Alone Dedicated Control Channel
SDK	Standard Development Kit
SDR	Software Defined Radios
SGBV	Sexual and Gender-Based Violence

SIDA	Swedish International Development Cooperation Agency
SIM	Subscriber Identity Module
SIP	Session Initiation Protocol
SMPP	Short Message Peer-to-Peer
SMS	Short Message Service
SMSC	Short Message Service Center
SOP	Standard Operating Procedure
SQL	Structured Query Language
SS7	Signaling System 7
SSHRC	Social Science and Humanities Research Council of Canada
TU WIEN	Vienna University of Technology
UDHR	Universal Declaration of Human Rights
UK	United Kingdom
UN OCHA	United Nations Office for the Coordination of Humanitarian Affairs
UN OIOS	United Nations Office of International Oversight Services
UN OOSA	United Nations Office for Outer Space Affairs
UN	United Nations
UNDAC	United Nations Disaster Assessment and Coordination
UNESCO	United Nations Education, Scientific, and Cultural Organization
UNFPA	United Nations Population Fund
UNHCR	United Nations High Commissioner for Refugees
UNICEF	United Nations Children's Fund
USAID	United States Agency for International Development
US-VISIT	United States Visitor and Immigrant Status Indicator Technology
VoIP	Voice over Internet Protocol
VoLTE	Voice over LTE
WASH	Water, Sanitation, and Hygiene
WFP	World Food Program
WHO	World Health Organization

1 Introduction

Carleen F. Maitland

In the spring of 1992, while serving as a Peace Corps volunteer on the border in Malawi, I was thrust into a humanitarian crisis resulting from the Mozambican civil war. As a new wave of refugees flowed over the border, I joined Médecins Sans Frontières staff, assisting with registration. Witnessing the hardships and trauma suffered by those approaching the wooden table in the middle of the woods, as well as ongoing interactions with local camp residents and staff, I gained a keen awareness of their vulnerability. More than twenty years later, as the refugee crisis of the early 21st century continues to unfold, I once again find myself addressing these vulnerabilities.

Even though the current crisis has similar roots in armed conflict and persecution, it plays out against the backdrop of a technical revolution and a transformative era in humanitarian operations. As mobile phones, remote sensors, biometric technologies, and wireless networks proliferate around the globe, the humanitarian community has begun to embrace these new technologies. It has been assisted in these endeavors by "hacktivists," volunteer technical communities, and refugees themselves.

Together, these forces of innovation represent enormous potential for improving lives, but they can bring harm as well. Whether or not long-term benefits of new information and communication technologies (ICTs) are realized will depend, in part, on their sustainability and scalability. This volume contributes to these longer-term goals by reflecting on current developments, while also looking to the future.

The authors, from disciplines as diverse as international law, international affairs, and information and computer science, provide knowledge and expertise on forced migration, technical design and development, and user behavior. Together, they provide unique insights into the operations and systems being used for refugee support, which range from forecasting refugee flows to providing more direct assistance to the affected population, including refugees' use of technology. Their contributions embrace new technologies while voicing concerns about ubiquitous and uncontrollable data flows.

Scalable and sustainable ICTs must be effective in a range of contexts, considering the global nature of the crisis. The recent influx of migrants into Europe from a variety of countries has captured headlines. However, the countries closest to the conflicts host the largest populations of refugees, with the top ten primarily located in the Middle East (Turkey, Pakistan, Lebanon, Iran, Jordan) and Africa (Ethiopia, Kenya, Uganda, Chad). The exception is Germany, ranked number nine (UNHCR, 2017). Yet, as the Rohingya crisis in Mayanmar suggests, these rankings are subject to change.

The insights presented in these pages reflect this diversity, derived from the authors' direct field research and collaborative projects conducted in a wide range of countries, including Jordan, Lebanon, Rwanda, Germany, Finland, Sweden, Norway, Italy, and Greece, as well as the United States and Canada. They also contribute insights developed through participation in international forums and service to international organizations.

Nevertheless, for ICT innovation, the European crisis, with its widespread network infrastructure, highly trained IT workforce, and wealth, has generated many new startups, providing innovative services and technologies. While many target refugees in Europe, others have focused on more distant locales as well. With similar ventures emerging in Jordan, Turkey, and Kenya, among others, the desire to help through technology can be characterized as truly global.

These startups join established technology firms' Corporate Social Responsibility programs to provide relief while driving innovation. As revealed throughout this volume's chapters, firms as diverse as Microsoft, Google, and Facebook, among others, have not only contributed staff and existing technology, but also have worked to develop new tailored applications and approaches.

Whether or not these innovations from startups or established technology firms achieve sustainability, they provide immediate benefits. Moreover, their general concepts can serve not only as a bellwether for future humanitarian technologies, but also as new means of organizing support. As experienced tech industry and humanitarian staff observe, the long-term success of these exciting innovations and emerging startups will likely depend on effective coordination with networks of established humanitarian organizations. As discussed in this volume's final chapter, these interactions, coupled with rapid technological change, portend a less centralized, more nimble and innovative approach to assisting the emerging *digital refugee* through systems of *digital humanitarian brokerage*.

The ensuing pages lay the groundwork for these concepts, starting with an overview of trends in the humanitarian community. This is followed by further discussion of the book's framing and underlying themes, as well as the information needs of refugees and organizations. The chapter then concludes with a description of the volume's structure.

Trends

The broader humanitarian sector's recent embrace of innovation is easily visible in changes at the UN. There, agencies are following the path charted by UNICEF, which began its formal innovation efforts in 2007 (Jolly, 2014; UNICEF, 2014). At sister organization UNHCR—the UN Refugee Agency—a similar dedicated "Innovation Unit" was launched in 2012, to identify, formalize, and promote innovations within the agency. Following UNICEF's model, UNHCR launched a staff fellows program,[1] as well as a series of geographically distributed labs designed to foster and promote technical innovation. The fellows program aims to amplify field-based innovation, such as UNHCR's early use of biometrics in the 2003 Afghan repatriation program. As compared to the fellows program, the labs are more externally facing, leveraging innovations from both within and outside the organization. One of four labs, the Link Lab specifically focuses on the use of ICTs for refugees, as well as information sharing among partners.

As is generally recognized in innovation studies, often times technological change is driven by grassroots efforts. Not surprisingly, international and local non-governmental organizations (NGOs) are also making more extensive use of and innovating with ICTs as well. One example is NetHope, a consortium of international NGOs, formed in 2001 to promote and coordinate ICT use throughout its members. Having expanded from an initial 7 to 42 members, it now undertakes its own programs.

Over the last two decades, increased use of ICTs, lowered costs, and greater expertise and knowledge of both staff and beneficiaries has changed humanitarian organizations' management of these assets. The changing perceptions of donors also facilitated this trend. Long considered an element of overhead, ICTs are increasingly transitioning from the responsibility and budgets of central IT to those of individual programs. As a result, rather than being associated with email and web servers, they are considered costs attributable to essential programs, such as shelter, food, and sanitation. Seen as more critical activities within these decentralized budgets, greater resources and innovation are possible.

Alongside these organizational and managerial changes, are sectoral changes associated with global humanitarian reform. The reform movement views the current humanitarian system as fundamentally flawed, underpinned by outdated colonial frames, and not fully embracing new technologies and means of organizing assistance. Its leaders argue for a greater emphasis on involvement of local NGOs, diversification in donor countries, which currently are predominantly in the West, and a greater recognition of, and perhaps integration of, "new humanitarians" (Barnett & Walker, 2015; Sezgin & Dijkzeul, 2016).

In May 2016, the UN Secretary General convened the World Humanitarian Summit to address the unprecedented humanitarian challenges around the globe. The Summit emphasized the need for a greater orientation toward mitigation rather than response, greater focus on local and regional organizations in these activities, and the promotion of innovation. During the meeting, the Global Alliance for Humanitarian Innovation was formed to "accelerate transformative improvements in humanitarian action by creating a shared space for the development of innovative tools, approaches and processes."[2]

Framing

Amid this backdrop of technological, organizational, and sectoral innovation, this volume's structure is aligned with sociotechnical systems theory, capturing the co-evolution of the above-described social and technological systems of displacement. The theory's tenets emphasize the duality of social and technical systems, and importantly that technological outcomes are unknown, may be positive, negative, *or both*. Sociotechnical systems theory also views these systems as separate, yet mutually dependent. Accordingly, the volume is organized into social and technical sections, followed by syntheses reflected in a chapter devoted to the policy dimensions of these changes, and a conclusion section identifying and specifying research directions on emerging trends.

Collectively, the volume builds on recent findings on ICTs in forced migration, focusing on current and emerging uses, and extrapolating to generate future-oriented research questions. The digital lifeline metaphor introduced in the title, particularly its being posed as a question, reflect both the volume's critical nature as well as the stance that questions of technology's impacts for refugees are not settled. As technology changes, questions of whether and how they affect the lives of the displaced must be asked and analyzed anew.

Embedded in the sociotechnical framing are underlying themes of networks, mobility, and connectivity as well as the refugee lifecycle. Network structures pervade technical, interorganizational, and interpersonal relations. Whether in the social networks, networks of organizations, or the layered, modular architectures of wireless, cellular, and fixed networks, these structures support gathering, sharing, aggregating, and processing data.

There is a dichotomy between the mobility of persons and their mobile devices that have become indispensable to so many, and the telecommunication networks that are less so. A constant search for connectivity ensues, during flight, but also after reaching a place of asylum. Connectivity can be hampered by lack of infrastructure or mere poverty and lack of economic opportunities. In many cases, the extent to which mobility, networks, and connectivity provide benefit or harm are determined by information policies—at the international, national, and organizational levels.

The lifeline metaphor also reveals this volume's orientation toward organizational perspectives. While many of the authors are directly engaged with the displaced, we view organizations as critical to long-term technical innovation, whether as the providers of new, innovative applications, networks, and services used by refugees and organizations, or the entities leveraging their use. Again, the metaphor of the lifeline reflects the position of the relatively safe and secure (organization) extending a "lifeline" to someone in need.

Finally, our analyses need to be understood in relation to the overall experience of displacement. While often extremely complicated, with recursive periods of flight and temporary settlement, here, in a highly simplified form, the process is described as the "refugee lifecycle." Viewed as a cycle, stages can be defined by the physical and legal context, each with unique information needs and contexts influencing their fulfillment. The lifecycle is also helpful for situating and identifying gaps in ICTs for forced migration scholarship.

Among the cycle's four stages, the first is *departure and transit*, occurring when conditions deteriorate to the point a person or family is forced to flee. The journey to safety may end in a neighboring town or province, at the first international border, or may also include transiting several countries before reaching the chosen destination. The second stage for some, but not all, is *legal determination*, where, if appropriate, the migrant seeks formal recognition of their refugee status by applying to appropriate authorities, and, once determined, registration follows. In the third stage, *temporary asylum*, the displaced establish temporary residence, typically in an urban or camp environment. Finally, in the fourth stage, *permanency*, the displaced are either: (1) locally integrated into the country of first asylum, being offered a permanent legal status; (2) resettled to a third country; or (3) repatriated to their home country. The chapters in this volume focus primarily on the second and third stages of legal determination and temporary asylum.

Information Needs

The lifecycle model can be helpful in identifying the wide ranging and changing information needs, often, but not exclusively, fulfilled through the use of ICTs, by refugees and humanitarian organizations alike. At each stage, refugees and the organizations from which they receive support, operate in frequently changing networks. For the organizations, in each country and each crisis, networks consist of different agencies, whose participation depends on a variety of factors. These factors include proximity to the crisis, community-based versus regional versus international operating footprint, sectoral focus, and the stage of the crisis or lifecycle of the displaced.

These complex refugee service networks consist of organizations ranging from UN agencies, governmental organizations—whether representing the government of the crisis area or the hosting nation—and a wide range of international and local NGOs. While governments maintain sovereignty rights, in many low-income countries, where the majority of refugees are found, UNHCR typically serves as the response coordinator.

This coordination role is specified by the UN's Cluster Approach to crisis coordination, which is overseen by the Inter-Agency Standing Committee (IASC). In this system, each UN agency (and sometimes international NGOs) are identified as "cluster leads," responsible for response coordination.[3] The clusters include food and security, health, logistics, nutrition, protection, shelter, water/sanitation/hygiene (WASH), camp coordination and management, early recovery, education, and emergency telecommunications.

Important for the work here, cluster leads are responsible for *intra-cluster* information management, while the UN Office for the Coordination of Humanitarian Affairs (UN OCHA) is responsible for *inter-cluster* sharing (IASC, 2006). The World Food Program's role as cluster lead for humanitarian logistics as well as emergency telecommunications situates it as a locus of ICT innovation for the overall system.

Within the more narrow realm of refugees, UNHCR's information management responsibilities often stem from its data collection efforts during the registration process, part of the aforementioned overall legal determination stage. With a refugee registration identification number, much of the data collected throughout the network of humanitarian organizations are tied to that number, similar to a national ID. However, given the numbers and diverse backgrounds of the organizations through which these data flow, it is foreseeable that organizational cultural differences can be a barrier to information sharing (Palttala, Boano, Lund, & Vos, 2012).

Unsurprisingly, in the lifecycle stage of departure and transit, due to security risks and other complications,[4] information is critical but hard to find and validate. International organizations may have difficulty accessing the internally displaced, and local organizations may struggle to garner resources to help those in need. Constraints can originate not only from the conflict itself, but also from national governments seeking to limit involvement of external forces in domestic conflicts, or from international organizations' unwillingness to risk staff lives by placing them in a conflict zone. Information flows for the displaced and those seeking to flee abroad are complex, and often informal and shared through personal networks. Information on safe routes and conditions at destinations can change constantly.

Once abroad, the determination and registration process also creates unique information needs for the displaced and humanitarian organizations alike. The displaced have pressing needs to stay in touch with those back home as well as to keep updated on the status of the general situation that drove them to leave. In entering a new country, they are likely to be thrust into the Refugee Status Determination (RSD) process. While officially the responsibility of nation states, the process may be delegated to UNHCR, or handled through a hybrid approach.[5] At a minimum, the displaced will be asked to present identity documents, but typically much more information is required. Local or international NGOs may be present to provide the displaced advice on rights, representation, and appeals (USCRI, 2013), and where asylum application processes are drawn out, provide daily support for shelter, food, and clothing.

Having achieved temporary asylum, and its security and short-term stability, accessing assistance for food, shelter, clothing, and other resources requires the displaced to engage with sometimes large and complex networks of humanitarian or social service organizations. These service providers may be host country government agencies, local NGOs, or international NGOs together with UN agencies. The last may have existing operations in the country, working on long-term development projects but able to redirect efforts and ICTs to serve the displaced. For the displaced, accessing assistance requires entering what is often a labyrinth of information systems, receiving and providing critical information all while attempting to recover from trauma and loss.

Lastly, during the final stage of identifying permanent solutions, the displaced may be required to provide even more information. In the cases of formal repatriation or resettlement, UNHCR and the International Organization for Migration (IOM), typically work with the host country, the government to which they are being resettled, or the government from which they fled. For example, in the case of repatriation to Sudan, the International Organization for Migration provided service coordination for returnees while at the same time serving as partner on the governing council, together with UNHCR and both the governments of Sudan and South Sudan (Weiss, 2012).

Across the various stages represented in table 1.1, refugees and service providers alike experience significant information needs. Fulfilling these needs requires innovative access and data management solutions. However, currently, access to networks and computing facilities tends to favor organizations, with the displaced often left to their own devices. Where cellular

Table 1.1
Refugees' and service providers' information challenges

Stage	Issues	Information needs of refugees	Information needs of service providers
Departure and transit	Security for the displaced and service providers; lack of legal status as refugees; IDPs vs. refugees.	Where to go? Why? When? How? With whom?	Where are people going? How? Why? When?
Determination and registration	Centralized provision, serves as a funnel for information; first step in process of providing support and mechanism for making referrals and tracking services.	Options? Pros and cons of each? Location of family members?	Identity? Nationality? Documentation? Records (school, health)?
Temporary asylum	Where? Why? How? Access to services? Learning entirely new system.	Extensive—how, where to live? Livelihoods? Resources?	Information about needs—education, physical and mental health services; livelihood opportunities.
Permanency	Where? How? Settling all over again.	Extensive—how? Where? When? Housing? Education? Support services?	Information about needs—education, physical and mental health services; livelihood opportunities.

network connectivity is available, refugees and the displaced may be able to maintain access, but mostly through their own means.

However, as recognition of the importance of connectivity has grown, greater efforts are being made to provide refugees with Internet access (UNHCR, 2016). Early efforts, such as those seeking to improve Internet back-haul capacity in rural camps like Dadaab, Kenya, aimed to benefit humanitarian organizations and refugees alike. More recent efforts, such as many providing temporary WiFi hotspot connectivity to the wave of migrants entering Europe, are directly targeting the displaced, providing a critical resource at a time of urgent need. However, as awareness of cybercrime and state surveillance grows, greater attention is being paid to securing these networks (Maitland & Bharania, 2017).

As ICTs continue to rapidly diffuse throughout refugee and humanitarian networks, questions arise as to the potential benefits and harms. Such questions include where technologies' benefits outweigh harms, how access and use should be managed and controlled, as well as supported. Each of the chapters in this volume addresses one of these elements. The authors, drawing on their extensive expertise and experience, imagine the future and pose further questions whose answers will provide important guidance. In so doing, they also offer a connection between scholarship from forced migration and information and computer science. Also, within these more technical disciplines, the research proposed herein can bridge the divide between Information and Communication Technologies for Development (ICTD) and Crisis Informatics (Maitland, Pogrebnyakov, & Van Gorp, 2006).

Outline of the Book

The goal of this volume is to provide an agenda for the fundamental social and technical research necessary to guide future innovation for improving the lives of the displaced. The cross fertilization of ideas among chapter authors from a variety of academic disciplines provides unique insights and approaches to pressing problems. Together, their collective expertise generates a ground breaking research plan with broad implications.

As the first of the book's three sections, the "Legal, Social, and Information Science Perspectives" presents research agendas from international law, as well as political, organization, and information science. In chapter 2, Galya Ruffer considers the informational aspects of the refugee status determination process, examining how technological change may affect such issues as testimony and records access. In particular, she addresses the question: How can ICTs contribute to a refugee status determination process that guarantees due process, access to justice, and ongoing protections?

In chapter 3, Lindsey Kingston considers identity documentation, statelessness, and displacement, and in particular how biometric technologies may influence the rights of the displaced. Building on the background in refugee status determination provided by Ruffer, and continuing through the stages of displacement to biometric use in registration and beyond,

Kingston explores ethical and legal issues surrounding biometric use. From there she raises research questions related to human rights and the potential for biometric data to be treated as property rights, with implications for ongoing access and control by the displaced.

The progression through the refugee lifecycle continues in chapter 4, which examines the role of multilevel governance in agencies providing support primarily during temporary asylum. Here attention is turned to the challenges that interorganizational contexts, including hierarchies and networks, create for information sharing. Consideration is given to how multiple levels of hierarchy and the structures, procedures, and policies they transmit, influence both technology adoption and information flows. The conclusion poses the questions: Where in an organizational hierarchy are information technology expertise and policies best developed to ensure both effective and efficient information sharing? And, how can networks of service providers most effectively deploy technologies and share information?

In chapter 5, the last in the "Legal, Social, and Information Science Perspectives" section, Karen Fisher sheds light on the information and technology use behaviors of migrants and refugees in transit, camps, and asylum. Her analysis provides a timely review of literature on the refugee crisis in Europe, and, in contrast to previous chapters, provides a greater focus on the issues for displaced individuals. Her chapter also reflects the firsthand views of refugees from the Za'atari Syrian refugee camp in Jordan and their perceptions of and hopes for technology.

The book's second section, "Technical Perspectives," examines technological change across a variety of levels. Similar to the levels of analysis spanned in the first section (international, national, organizational, individual), the chapters in this section span technical levels from network access, information systems, applications, and data analytics. The authors also represent different disciplines: computer science (networking and data), information systems, and geographic information science (GIScience).

In chapter 6, Elizabeth Belding and her team, led by Paul Schmitt, examine how future developments in cellular and wireless technologies may affect Internet access for refugees and service providers alike. Focused on camps, which are geographically limited and hence amenable to network implementations, they discuss innovations in cellular and wireless technologies that could help solve the access issues plaguing many camps. Their discussion addresses not only technical issues but also the fundamental reasons for insufficient bandwidth and access. These issues include economic factors, such as refugees' inability to afford broadband access, as well as policy issues related to some countries' restrictions on the spectrum required for wireless networks. Their research agenda highlights the importance of characteristics such as rapid deployability and ease of setup and breakdown, which accommodate temporary use, in future designs.

Next, chapter 7 examines a selection of humanitarian organizations' information systems, an understanding of which is critical for scalability and sustainability of new innovations. Through detailed descriptions of data capture and interorganizational data management systems, this chapter demonstrates the variety of challenges to control, implementation, and benefit, as well as questions they raise. The proposed research agenda includes questions related to conflict in systems deployment and use, the need for comparative analyses of technology deployments, and critical analyses of technologies' impacts on the beneficiaries of refugee service systems.

In chapter 8, Brian Tomaszewski continues the focus on information systems, examining the use of GIS and the role of spatial thinking in displacement, and the research needed to more effectively use both to improve the lives of the displaced. His approach integrates theories of human geography with the practical application of geographic information systems. Tomaszewski raises important questions related to refugees' conceptualizations of space. Torn from their homes and towns, and in a place of temporary asylum that may be changing all around them (particularly when camps are being set up), how do refugees make sense of space? In particular he poses the question: What role can GIS and maps play for multi-scale refugee situation awareness?

In the last chapter of the section, chapter 9, Susan Martin and Lisa Singh consider how new data analytic techniques might be used to improve prediction and analysis of forced migration and identify research directions leveraging "big data analytic" techniques for the forced migration community. Their contribution covers the methods for generating the data necessary to perform predictive modeling, which in turn can generate more timely and accurate warnings of forced migration events. Warnings are critical in helping service providers pre-position supplies and plan deployments. Their work is truly interdisciplinary.

The book's third and final section synthesizes the volume's contributions to identify policy research directions (chapter 10) as well as a unified research agenda reflecting two important trends, along with tips for implementation, drawn from extensive field experience (chapter 11). The former analyzes the social and technical research agendas to develop an associated information policy research agenda for both national and organizational levels. The agenda is unique in its breadth and global dimensions, combined with a grounding in the issues of forced migration. The agenda will establish fertile ground for academic information policy research with far reaching impact for this vulnerable population. For practitioners, the agenda may spur change within organizations where decision-makers have the power to implement changes.

The final chapter, chapter 11, presents a unified research agenda based on two emerging trends, the rise of the *digital refugee* and the increasing importance of *digital humanitarian brokerage* as driving forces in humanitarian service provision. Both have their roots in more ubiquitous connectivity as well as the increasingly central role for data management. Following this, I close the volume with practical advice, addressing challenges, both logistical and ethical, in conducting displacement research.

Notes

1. http://innovation.unhcr.org/fellowship.

2. http://www.thegahi.org.

3. https://www.humanitarianresponse.info/en/coordination/clusters/what-cluster-approach.

4. Complications include gaining access to those in need, permission to operate in the country, the potential for assistance to be used as support for either side to the conflict and consideration of whether limited resources might be more effectively deployed to support those who have fled to neighboring countries. For a discussion of these issues in the Syrian case, see http://world.time.com/2014/01/14/syrias-humanitarian-disaster-how-aid-has-become-a-weapon-of-war.

5. However, not all countries are signatories of the 1951 Convention nor the 1967 Protocol, and in some cases signatories lack the complementary national laws. As a result, these countries turn to UNHCR to conduct registration procedures or develop hybrid systems in which the host country is also involved.

References

Barnett, M., & Walker, P. (2015). Regime Change for Humanitarian Aid. *Foreign Affairs, 94,* 130–141.

Jolly, R. (2014). *UNICEF (United Nations Children's Fund): Global Governance that Works.* New York: Routledge.

IASC. (2006). *Guidance Note on Using the Cluster Approach to Strengthen Humanitarian Response.* United Nations. Retrieved from https://interagencystandingcommittee.org/system/files/legacy_files/Cluster%20implementation%2C%20Guidance%20Note%2C%20WG66%2C%2020061115-.pdf.

Maitland, C., & Bharania, R. (2017). Balancing Security and Other Requirements in Hastily Formed Networks: The Case of the Syrian Refugee Response. In *TPRC 45: 45th Research Conference on Communications, Information and Internet Policy.*

Maitland, C., Pogrebnyakov, N., & Van Gorp, A. F. (2006). A Fragile Link: Disaster Relief, ICTs and Development. In proceedings of the *2006 International Conference on Information and Communication Technology and Development, ICTD2006* (pp. 339–346). IEEE. Retrieved from doi:10.1109/ICTD.2006.301873.

Palttala, P., Boano, C., Lund, R., & Vos, M. (2012). Communication Gaps in Disaster Management: Perceptions by Experts from Governmental and Non-Governmental Organizations. *Journal of Contingencies and Crisis Management, 20*(1), 1–12. Retrieved from doi:10.1111/j.1468-5973.2011.00656.x.

Sezgin, Z., & Dijkzeul, D. (Eds.). (2016). *The New Humanitarians in International Practice: Emerging Actors and Contested Principles.* New York: Routledge.

UNHCR. (2016). *Connecting Refugees: How Internet and Mobile Connectivity Can Improve Refugee Well-Being and Transform Humanitarian Actions.* Retrieved from http://www.unhcr.org/5770d43c4.pdf.

UNHCR. (2017). *Mid-Year Trends 2016.* Retrieved from http://www.unhcr.org/58aa8f247.pdf.

UNICEF. (2014). *Annual Report.* Retrieved from https://www.unicef.org/supply/files/UNICEF_Supply_Annual_Report_2014_web.pdf.

USCRI. (2013). *Refugee Status Determination in Latin America: Regional Challenges and Opportunities.*

Weiss, T. (2012). The Transition from Post-Conflict Assistance to Rehabilitation in Sudan: An IOM Contribution to State-Building and Reconstruction. *Lund Horn of Africa Forum.*

I Legal, Social, and Information Science Perspectives

2 Informational Components of Refugee Status Determination

Galya Ben-Arieh Ruffer

Over 60 million people are displaced as a result of persecution, conflict, generalized violence, and human rights violations. The *1951 United Nations Convention Relating to the Status of Refugees* and *1967 Protocol* ("*Refugee Convention*") created a legal framework for refugees, defined as any person who, "owing to a well-founded fear of being persecuted for reasons of race, religion, nationality, membership of a particular social group or political opinion, is outside the country of his nationality, and is unable to, or owing to such fear, is unwilling to avail himself of the protection of that country."[1] States who are parties to one or both of these Conventions ("party states") are incorporated into a complex legal framework comprised of additional conventions such as the Convention Against Torture, the Convention on the Rights of the Child and the Convention on Statelessness, regional conventions and instruments, and national legislation that together form the international refugee regime. The cornerstone of the *Refugee Convention* is the refugee status determination process ("RSD"). While a refugee who is present on the territory of a state is protected by the principle of *non-refoulement* (that a refugee will not be sent back to a place where his life or freedom could be threatened), it is through the presentment to an official authority and a refugee status adjudication process that refugees can more fully access the rights states are mandated to provide under the *Refugee Convention*.

Access to refugee status is urgent and crucial to securing refugees access to the rights of refugee protection. At the same time, even wealthy states with the strongest refugee legislation in the world, such as South Africa, lack the capacity to process large-scale numbers of refugees, and, given political constraints, fail to offer rights based protection (Amit & Landau,

2016). Although some refugee scholars and practitioners have advocated that the office of the United Nations High Commissioner for Refugees (UNHCR) be granted the exclusive authority to conduct RSD with sufficient resources to do so, it is unlikely that states will cede the authority to grant refugee status to UNHCR or some other supranational authority (Frelick, 2016). Even if states were to cede the authority, practitioners who represent refugees seeking UNHCR recognition in places such as Egypt, Malaysia, and Thailand highlight concerns that centralizing RSD in UNHCR might not be desirable from the standpoint of refugee protection absent additional safeguards, such as a right to legal counsel, a formalized appeals process, and other procedural guarantees.

Given the overloaded refugee status determination process worldwide, what if we could harness the potential of information and communication technologies (ICTs) to transform the RSD process into a more efficient, reliable and more broadly accessible system that maintains national authority? Given the scale of resources currently used in RSD, a transformation in the use of ICTs in the RSD process could reduce costs, which would enable states and the UNHCR to shift resources into projects that further the overarching goal of the *Refugee Convention* protection regime, such as advancing the ability of refugees to become self-sufficient, reducing the burden on states through access to rights such as freedom of movement, education, and livelihoods.

Even absent the exclusive authority, the role of UNHCR in conducting RSD has grown over the past decade as border security in Western states and ongoing cycles of violence have resulted in refugee flows into new destination countries. Although many countries have signed the *Refugee Convention*, 31 state parties do not have implementing legislation resulting in the UNHCR being primarily responsible for conducting RSD. Part of UNHCR's work is to encourage countries to transition to a national RSD system. As a result, transitions have recently taken place or are underway in Israel, Korea, Ecuador, Japan, Mexico, Malta, Kenya, Morocco, Turkey, Hong Kong, and Cameroon. While these states have been transitioning to systems in which the national government conducts RSD through the implementation of a formal national process, UNHCR remains the second largest adjudicator of refugee status in the world (after South Africa) (UNHCR, 2014). In 2014 the UNHCR issued 134,500 decisions out of 233,500 applications, and in 2015 the number of decisions rose to 135,900 out of 257,600 applications received.[2] While in 2014 there was a backlog of 303,000 applications pending, this number increased to 460,000 in 2016. While some of these will not be found to be refugees, they still have to undergo an RSD adjudication process. With the Syrian refugee crisis, most of the 2016 cases are in Turkey where the national government will be taking over RSD.

Given the rising numbers of RSD, UNHCR has been considering the protection impacts of RSD and undergoing a number of shifts to reduce the resources invested in situations where RSD does not change much in the protection of individuals. The primary one is an increase in the use of group determinations on a prima facie basis for situations where the need to provide assistance is urgent and the number of people affected by the crisis make an individual determination of refugee status for each member of the group impractical. A prima facie approach means the recognition by a state or UNHCR of refugee status on the basis of readily apparent, objective circumstances in the country of origin.[3] While this idea has been put forth as a way for UNHCR to process refugees in countries where it has been either delegated or has assumed this responsibility, the information and communication technology (ICT) components that would enable greater reliance on prima facie recognition have not been sufficiently addressed from the operational standpoint of how refugees and practitioners would be able to access reliable data of country conditions or the legal issues a prima facie system raises for national obligations to refugee protection. The second shift is to increase the quality and efficiency of RSD when it is used through procedural standards, and the final shift is to work with emerging asylum systems to improve state procedures.

Three broad challenges for which ICTs can be considered in this new RSD context are: (1) meeting capacity in large-scale arrivals; (2) meeting capacity across a range of sites and countries; and (3) enabling mobility such that once a refugee has been recognized, she can reach a host country that can afford her protection. After introducing the context of RSD today and the role of prima facie recognition, this chapter offers an analysis of the importance of the recognition process in the *Refugee Convention* scheme, how ICTs could contribute to present challenges, and to fulfilling the vision of the *Refugee Convention* regime.

One of the main challenges today for refugees seeking protection is the need to establish identity and fear of return to their country of origin resulting from the face-to-face interview with a government official in order to be allowed entry and access to the RSD process. Through ICTs we can imagine a world in the future where a refugee does not need to meet physically with an official to pass this first hurdle in establishing a claim for refugee status. Two core evidentiary elements—identity and nationality—can be established through a mobile device. Momentarily suspending recognition of the complexities, one can envision a mobile device confirming, through location tracking, the exact spatial coordinates of the person, providing evidence of having crossed an international border as per the definition of a refugee under Article 1 of the *Refugee Convention*.[4]

A greater reliance on mobile technologies is already underway through government digitization programs as sites of state transformation. While technologies of mobility raise concerns of data tampering and loss, storage and complexities of data transfers across different systems, as well as unwanted surveillance, these need to be counterbalanced with the situation we have today wherein refugees, in desperate need of protection, are unable to access an RSD process at all. If we can create a more efficient, reliable, and accessible system to recognize refugees, the bigger questions that emerge concern transit across multiple state borders as a person seeks protection and access to rights. Therefore, we can also consider how ICTs might contribute to the ability of a refugee to access an official international travel document, similar conceptually to the historic Nansen passport, that would enable a refugee to travel across state borders in order to reach a country that is willing and able to provide protection. Finally, it tends to be the case that states more readily granting access to their territory, and, therefore, housing most of the refugees in the world today, are also the ones restricting access to rights given a lack of resources. Therefore, if we are able to harness the potential of ICTs to achieve mobility, where refugees and forced migrants can more easily be granted recognition, the challenge then becomes how to allocate resources based on protection needs and vulnerabilities. In other words, how can ICTs enhance burden sharing and solidarity?

Perhaps we can imagine a world where a refugee, once officially recognized through a mobile device, would be directed to a central matching process that, according to preferences and other criteria, would arrive at a selection list of possible hosting states. The questions in achieving this longer-term vision are: (1) What would nation-states agree to if costs were reduced? (2) How can we devise such a preference-based system (and what criteria should be included)? Whether we envision a world where a more fully developed form of prima facie recognition would replace the RSD process or a world of more secure RSD, we need to consider: (1) how to create portable identity that is protected and secure; (2) how to connect status to rights; and (3) how to better enable burden sharing among states.

The Context of RSD Today and the Role of Prima Facie Recognition

Around the world today, states and the UNHCR are struggling to cope with refugee processing. Consider the situation in Greece where the European Commission has recommended that Greece ensure the Asylum Service has additional staff to process the intake of asylum applications, to detect bottlenecks, and to guarantee that cases are handled in a well-organized manner, and that there are adequate administrative support staff to book interviews, make appointments for registration, book interpreters, and manage workflow of each office for a well-functioning asylum process (European Commission, 2016, p. 10). In order to respond to the increase flow of asylum seekers into Greece, the Commission recommended the establishment of an increased number of permanent open reception facilities to correspond to the number of applicants for international protection and the capacity to open temporary facilities on short notice with UNHCR contracted to provide free legal assistance to applicants at the appeals stage.

The challenge is that the registration of refugees and the intake of asylum applications are thought of as a face-to-face activity through which refugees present themselves to an official in order to be awarded. This thinking persists even as ICTs play an increasingly greater role in other aspects of the RSD process, and, more broadly, in government digitization programs. For example, Central American refugees in U.S. detention centers appear before immigration judges through teleconferencing and interpreters are available through call in systems. Within consular processing, applicants for an entry visa are prompted to upload identity documents, a photo, signed consular forms, and other required proof through their mobile phones, through which they then schedule an appointment. Given these developments, it's important to first understand the relationship between official recognition by a state or UNHCR and a person's refugee status before we consider the question or whether and how a shift to prima facie recognition might adapt to new processes that can harness the potential of ICTs in the prima facie recognition process.

States have the primary responsibility for determining the status of asylum-seekers, but UNHCR may do so where states are unable or unwilling. As states in the Global South, and, increasingly, in the Global North find the influx of refugees beyond their institutional capacity, UNHCR has taken up the role. Unlike states that must enact legislation to create the administrative apparatus to adjudicate refugee status, UNHCR processes refugees under its *Mandate Statute*.[5] The core standards, including reception and registration, are presented in the *Procedural Standards* for RSD under UNHCR's Mandate.[6] With record numbers of people displaced in the world and stretched resources, UNHCR has been revising the Procedural Standards as part of a new strategic direction for RSD moving toward greater use of prima facie recognition, alternatives to RSD, and new ways of identifying people most in need of protection.

Group determination of status on a prima facie basis is a mechanism through which a state or UNHCR recognizes refugees in need of protection on the basis of objective circumstances in the country of origin. In the literature, it is often associated with exceptional situations in which rapid arrival or large numbers of asylum-seekers may overwhelm the state's capacity to implement an individual administration of their claims (Rutinwa, 2002, p. 3). It has also, however, been part of the Canadian refugee recognition process in which refugees with manifestly clear claims undergo a fast-track process, leaving the more complicated cases to a lengthier individual status determination process.

The shift to increased UNHCR RSD and prima facie recognition is in line with James Hathaway's proposal for global reform that recommends a shift away from national, and toward international, administration of refugee protection. The Hathaway proposal advocates a revitalized UNHCR to administer quotas, with authority to allocate funds and refugees based on respect for legal norms (Hathaway, 2016). In Hathaway's proposed RSD model, UNHCR would be expanded to assume the responsibility for a common international refugee status determination system and group prima facie assessment to reduce processing costs, thereby freeing up funds for real and dependable support to frontline receiving countries—including startup funds for economic development that links refugees to their host communities and that facilitates respect for the Convention rights of refugees in their host country or their eventual return home (Hathaway, 2016).

The reasoning behind the call for greater prima facie recognition is that, as established in the *Refugee Convention*, the role of the state is merely to recognize a refugee who enters the territory. According to the UNHCR *Handbook and Guidelines on Procedures and Criteria for Determining Refugee Status under the 1951 Refugee Convention and the 1967 Protocol relating to the Status of Refugee* (hereafter "UNHCR, *Handbook*"), recognition of refugee status does not make a person a refugee, but rather declares her to be one.[7] In other words, a person is a refugee as soon as she meets the criteria in the *Refugee Convention*. The refugee status recognition process, therefore, is a declarative one and not constitutive of the grant of refugee status. The role of the adjudicator is to confirm that the person meets the criteria; the onus is on the state to contest the claim for refugee status, and a state cannot avoid the obligation by simply not having a hearing (Hathaway & Foster, 2014, pp. 25–26).

In conjunction with the declaratory process, Article 31 of the *Refugee Convention* provides for the non-penalization of refugees by Contracting States who are obligated to "not impose penalties, on account of their illegal entry or presence, on refugees who, coming directly from a territory where their life or freedom was threatened in the sense of Article 1, enter or are present in their territory without authorization, provided they present themselves without delay to the authorities and show good cause for their illegal entry or presence," and that "Contracting States shall not apply to the movements of such refugees restrictions other than those which are necessary and such restrictions shall only be applied until their status in the country is regularized or they obtain admission into another country." Indeed, the *Refugee Convention* was a specific response to the recognition that individuals fleeing persecution from German-controlled countries during WWII were often denied admission to the United States and other countries because they had to cross borders without the requisite visa or documentation. The combination of the declaratory process of refugee recognition and Article 31 non-penalization on account of illegal entry recognizes the imbalance of power in that a refugee cannot force a hearing (Hathaway & Foster, 2014, pp. 28–30). The drafters placed the onus on the state to contest that the person is a refugee exactly because, otherwise, it would be too easy for a state to circumvent its obligation to protect refugees by simply never asking the question "are you a refugee?"—and thereby avoiding constituting that a person *is* a refugee (Hathaway & Foster, 2014, p. 26).

As UNHCR looks to cope with the increased processing needs, the objective is to create a system that better comports with a declaratory process. Under this system the process would be streamlined for refugees who fled from states where the country conditions warrant a group prima facie determination that people from these states are refugees, meaning that no further evidence other than proof the refugee is from this state is generally required.[8] Although there is no official list, "[t]he status of persons found to be prima facie refugees is that they are presumptively refugees within the meaning of the relevant instruments. This presumption is conclusive unless it is dislodged by evidence that either any person was wrongly recogni[z]ed as a refugee or was liable to *exclusion* under the provisions of refugee law."[9] As a logical consequence of this, prima facie refugees are entitled to enjoy all the rights of refugees under the *Refugee Convention* and any other instrument applicable to them (Rutinwa, 2002, p. 4).

According to UNHCR guidelines, there is no difference between a refugee granted status through a prima facie process versus a refugee who undergoes an individual determination.[10] The problem is that, in practice, states have used this loophole to create temporary status, to otherwise limit the rights of refugees, or to subject refugees to an additional process.[11] The state practice of limiting the rights of prima facie refugees is in line with one school of thought with regard to the nature of the status that refugees recognized on a prima facie basis acquire (Rutinwa, 2002, p. 5). This school maintains that the prima facie concept is only a provisional consideration of a group of persons as refugees by UNHCR, and does not, in turn, require the state to complete the refugee status determination formalities to establish definitively the qualification or not of each individual (Rutinwa, 2002, p. 5). In other words, it's only a presumption of refugee status and is a device for preliminary decision-making on what is the separate question of refugee status. UNHCR's position, on the other hand, is that the determination that a group is prima facie a refugee group raises a presumption that the individual members of the group are refugees. As such, they can benefit from the international protection and assistance extended to them by UNHCR, on behalf of the international community. They retain their refugee character unless there are strong indications that they are not—or are no longer—to be considered as refugees (Rutinwa, 2002, p. 6; Jackson, 1999, p. 4). Rutinwa concludes from this literature that "[e]ssentially, the question which both schools are trying to answer is what is the probative effect or worth of a prima facie determination on the question of the status of the persons concerned under refugee law?" In other words, as UNHCR is now asking as it seeks a new vision for RSD, what "added value" does the individual status determination provide? In that streamlining RSD through a prima facie group status determination might provide an opening through which states can avoid their *Refugee Convention* obligations by granting only temporary status, can ICTs provide the necessary streamlining that would provide the efficiency of a prima facie process as an individualized process? This would then close the loophole presented in these two schools of thought.

In that the question turns on proof, ICTs have an evidentiary role to play. Rutinwa argues that since prima facie is a rule of evidence—meaning "on its face"— that the objective evidence offered is sufficient to establish that the person is a refugee in a group determination system. In posing the question of how the rule of evidentiary burden applies, particularly when the issue is whether or not an asylum seeker who arrives as part of a mass influx is a refugee, Rutinwa argues:

> As under the general rules of evidence, the burden of proving a claim of refugee status lies with the person who submits that claim. However, due to the peculiar circumstances in which asylum seekers find themselves, the duty to ascertain and evaluate the relevant facts is, under refugee law, shared between the applicant and the examiner. This then means that in principle the evidential burden rests, in the first instance, with the claimant. Where the claimant discharges that duty, either by themselves or with the assistance of the examiner, then s/he is presumed to be a refugee within the meaning of any applicable definition and the burden to disprove this provisionally shift to any person who opposes this view. If for any reason no such evidence is adduced, the evidence by, or in favour of, the claimant becomes conclusive." (Rutinwa, 2002, p. 8)

In a shift to a prima facie system, once the claimant submits the informational components—proof of country of origin and identity documents—the claimant's evidentiary duty would be discharged, and the presumption would be that she is a refugee, unless proven otherwise by the examiner.

The second issue concerns the information related to the country of origin. This information is needed, for example, to rebut a presumption by UNHCR that the country of origin is "safe," in other words, is not a refugee-producing state. UNHCR seeks to have a streamlined process through a centralized country of origin documentation center that would classify countries. For a claimant coming from a country presumed "safe" by UNHCR, status would be prima facie denied and only through an appeal would a person receive an individualized adjudication of her claim. The idea in streamlining the so-called "clear" cases is that the individual refugee status adjudication process would be a much smaller caseload of people who fled from states where the country conditions are less clear cut. In that this system hinges on country of origin documentation, it's important to note that over the past decade there has been a proliferation in national country of origin documentation centers. This trend needs to be further examined through systematic studies (Lawrance & Ruffer, 2015).

If we look at UNHCR's objectives in shifting to a prima facie approach from the perspective of the informational components that would be needed to support a declaratory process, we can see that a real-time assessment of protection and the ability to grant rights quickly in situations of mass influx enhance fairness. It would also enable better consistency in the refugee status adjudication process. However, fairness and consistency depend on reliable information on country conditions that are readily accessible and sufficiently detailed to enable an informed decision.[12] These informational components of a prima facie approach are mainly thought of in situations of mass arrival, but can also be used "in relation to groups of similarly situated individuals whose arrival is not on a large-scale, but who share a readily apparent common risk of harm" based, for example, on their ethnicity, place of former habitual residence, religion, gender, political background or age."[13]

On the other hand, in that the current challenge of a prima facie approach is that the main countries hosting refugees today feel overburdened, leading them to deny rights to refugees, if ICTs are able to fill the informational components and large numbers of people are recognized as refugees, in what way can ICTs further contribute to distributing refugees equitably among countries or enabling refugees to choose where they want to go in a way that better connects with local capacity to receive them? In that group recognition results in the problem of over recognition of some refugees and under recognition of others because of the information gaps (excluding manifestly unfounded claims), further research is needed to address the question of whether ICTs can somehow sort out people who do not need or deserve protection so that states do not become overburdened.

Finally, there are informational components in the determination to end the group prima facie designation. According to UNHCR guidelines, "[t]he decision to adopt a prima facie approach, therefore, needs to be kept under periodic review, such that the on-going use of the practice is deliberative and that security concerns are taken into account.[14] Likewise, through registration, the profile of individuals and their reasons for flight can be monitored on a continual basis."[15,16] The need for periodic review of country conditions and monitoring of the reasons people are fleeing in order to decide whether there is a basis on which to continue the prima facie process presumes the ability to maintain a reliable database, with secure profiles of individuals that can provide an accurate analysis of security concerns.

To summarize, the recognition of refugees is crucial and forms the basis for triggering states' rights obligations to a person as a refugee under the *Refugee Convention*. Given that refugee status is a declaratory process, recognition is an identification process and does not make a person a refugee. As such, when a person crosses an international border, it is important that they can present themselves to an official and that a system exists through which they can easily be identified as a refugee and afforded status, and, thereby, rights in accordance with the law. Under the *Refugee Convention*, a perfect system would enable a person to access refugee status as soon as she becomes a refugee under the definition. In other words, it would link up the status with the individual as opposed to the recognizing country. And, it would enable this recognition to happen virtually instead of requiring physical official contact with the refugee. The role of the states that are party to the Convention would be to grant the rights that refugees are owed under the Convention.

Reliability, Accuracy, and Accessibility of RSD, and the Role of ICTs in Prima Facie and Individualized Recognition

Given the crucial role of recognition, whether in a prima facie or individualized process, what is needed to have a reliable and accurate system that is broadly accessible?

Much of the focus on reliability and accuracy has been on country of origin documentation and the role of expert testimony (Lawrance & Ruffer,

2015). According to UNHCR Guidelines, country information should be relevant, current, and from reliable sources.[17] One of the main ways introduced to create consistency across RSD is through the development of a centralized country of origin documentation center. In recent years developments in the field have transformed country of origin information ("COI") introducing standards of accountability, based on the principles of relevance, reliability, backup, timeliness, and authority (Rusu, 2007; Hathaway & Foster, 2014, pp. 122–128). Additional principles that COI should be in the public domain, protect sources and personal data, guarantee impartiality (neutrality), and support transparency and accessibility (Rusu 2007, p. 7) are intended to enable COI to be considered in a full and value-neutral way. The belief is that there is no public interest, nor any legitimate individual interest in multiple examinations of the state of the backdrop at any particular time. Such revisits give rise to the risk, and perhaps even the likelihood, of inconsistent results. In addition, the standards intended to enhance consistency must be distinguished from deeming certain countries "safe" for purposes of adjudicating a refugee status claim. In other words, when a system relies on country of origin information to create a streamlined process for applicants from states with strong human rights records (i.e., "safe countries") in which the burden is shifted to applicants to rebut a presumption against well-founded fear is legally permissible—at least insofar as a fair hearing on the merits is provided—there is no legal basis for the wholesale exclusion of applicants from any given country (Hathaway & Foster, 2014, p. 127). In addition, there remain concerns in regards to the production of knowledge where UNHCR, which is beholden to states, produces COI in a centralized system. Finally, while UNHCR often has firsthand knowledge given its work on the ground, there are many situations where there is scant or conflicting information raising the question of whether ICTs can be used to crowdsource information.

A final informational component in the refugee status determination process that must be considered is the role of judicial and administrative assistance in supporting a reliable and accurate system of individual refugee status adjudication. Within the *Refugee Convention* itself, Articles 16 and 25 provide the rights of judicial and administrative assistance and establish the basis for accuracy. Judicial assistance may be of greatest value to recognition of status and "a fair assessment of refugee status is the indispensible means by which to vindicate Convention rights" (Hathaway 2005, pp. 630–631, pp. 634–635). Judicial and administrative assistance depends on being able to identify and connect people when they arrive in a country with the proper authorities and legal counsel. It also depends on government agents and local authorities having knowledge of the proper recognition and rights of refugees.

The Future of ICTs in the RSD Process

This final section raises questions around the potential use of technologies to support a streamlined process of refugee status determination. The discussion of the informational components of RSD has shown the potential for a family of integrated technologies that can more efficiently provide the requisite identity documentation including facial recognition, GIS location tracking, and translation of certificates, and a separate, but integrated family of technologies that can more efficiently and reliably be used to document the grounds of asylum, including video-based testimony, search engines, and linked databases of country of origin information. One question is whether already existing mobile technology, such as that used by the modern health system, where individuals can make online appointments, submit and verify insurance and other documentation, and receive lab results or consult with a doctor and nurse, is a model that can be used for the RSD process to enable a refugee to present identity documents to an official without having to be physically present? New technology would still be required to enable officials to take a person's biometrics through mobile technology and issue refugee status documents and travel documents.

Even if we can use and develop such integrated technologies, the larger questions concern the very different situation of a refugee who, unlike the user of a modern health system, lacks the protection of a state. It might be better to draw on models in use by refugees themselves and service providers who work with refugees in precarious situations to share and access information. For example, while mobile technologies already provide refugees with cash assistance and access to services, once a person presents herself to UNHCR or national authorities, is there a way to integrate these technologies with a status verification process that would not require the refugee to physically present herself to an official before receiving protection? This would go a long way in combatting dangerous onward movement and the use of smugglers. The creation of such a portable identity raises questions of property that Lindsey Kingston addresses in her chapter in this volume.

Where refugees do not have access to their documents or lack the ability to provide the necessary documentation, legal aid can be a matter of life or death. How can cloud-based technologies better connect legal aid providers, enable popup and mobile legal clinics, and support the development of strategic litigation and web training needed to create a pro bono global refugee legal aid corps on a global scale?

Whether we envision a world where a national refugee status determination process is no longer needed or one in which national RSD is simplified, the main point is how ICTs can contribute to a vision of human rights that reinforces the inalienable rights of the individual across national borders.

Notes

1. UNHCR, Handbook and Guidelines on Procedures and Criteria for Determining Refugee Status Under the 1951 Convention and the 1967 Protocol relating to the Status of Refugees, reissued, Geneva, December 2011, HCR/1P/4/ENG/REV.3 (hereafter "UNHCR, *Handbook*"), Annex II.

2. Author notes, "UNHCR RSD Retreat," Geneva, June 13, 2016.

3. UNHCR, Guidelines on International Protection NO. 11: Prima Facie Recognition of Refugee Status, 5 June 2015 (hereafter "*UNHCR, Prima Facie Guidelines*") available at: www.unhcr.org/558a62299.pdf, para. 1.

4. UNHCR, *Handbook*, Annex II.

5. UNHCR, Statute of the Office of the United Nations High Commissioner for Refugees, General Assembly Resolution 428(V) of 14 December 1950.

6. UNHCR, Procedural Standards for Refugee Status Determination under UNHCR's Mandate, Geneva, September 2005 (hereafter "UNHCR, *Procedural Standards*"), available at: http://www.unhcr.org/4317223c9.pdf.

7. UNHCR, *Handbook*, General Principles, para. 28.

8. *UNHCR, Prima Facie Guidelines*, para. 35: "Where there are indications of evidence to the contrary, persons need to be referred to a more enhanced registration process to gather more information. Where questions remain, the individual needs to be referred to regular refugee status determination procedures to assess adequately issues such as credibility and/or exclusion. In the event that regular status determination procedures are not operational, an assessment of the contrary evidence may need to be delayed, while making sure that the information is clearly recorded

within the registration system. This will have the benefit of facilitating a review of eligibility for refugee status or possible cancellation at a later stage, when individual processing becomes feasible and/or operational. In the meantime, such persons should benefit from an alternative form of stay."

9. UNHCR, *Prima Facie Guidelines*, para. 6: "A prima facie approach operates only to recognize refugee status. Decisions to reject require an individual assessment."

10. UNHCR *Prima Facie Guidelines*, para. 7: "Each refugee recognized on a prima facie basis benefits from refugee status in the country where such recognition is made, and enjoys the rights contained in the applicable convention/instrument. Prima facie recognition of refugee status is not to be confused with an interim or provisional status, pending subsequent confirmation. Rather, once refugee status has been determined on a prima facie basis, it remains valid in that country unless the conditions for cessation11 are met, or their status is otherwise cancelled or revoked."

11. UNHCR, *Prima Facie Guidelines*, para. 27: "In certain scenarios, it may be appropriate to apply a temporary protection or stay arrangement, as a prelude to a prima facie approach or at its end, even in state parties to the relevant instruments. In fluid or transitional contexts, such as at the beginning of a crisis where the exact cause and character of the movement is uncertain and hence a decision on prima facie recognition cannot be taken immediately, or at the end of a crisis, when the motivation for ongoing departures may need further assessment, a temporary protection or stay arrangement could be the appropriate response."

12. UNHCR, *Prima Facie Guidelines*, para. 41 addresses when prima facie is used within individual procedures, stating: "Adopting a prima facie approach in individual procedures has many advantages, not least those of fairness and efficiency. In terms of fairness, it allows like cases to be treated alike as far as decision-makers are required to accept certain objective facts relating to the risks present in the country of origin or former habitual residence. In terms of efficiency, such an approach would generally reduce the time needed to hear cases because individuals are required to establish only that he or she (i) is a national of the country of origin or, in the case of stateless asylum-seekers, a former habitual resident, (ii) belongs to the identified group, and/or (iii) the specified time period of the event/situation in question."

13. UNHCR, *Prima Facie Guidelines*, para. 10: "A prima facie approach may also be appropriate in relation to groups of similarly situated individuals whose arrival is not on a large-scale, but who share a readily apparent common risk of harm. The characteristics shared by the similarly situated individuals may be, for example, their ethnicity, place of former habitual residence, religion, gender, political background or age, or a combination thereof, which exposes them to risk."

14. UNHCR, *Prima Facie Guidelines*, para. 12 addresses security concerns under Art. 1(F) of the *Refugee Convention*: "A prima facie approach may not be appropriate in

all of the aforementioned situations, taking into account security, legal or operational factors. Alternative protection responses may be more suited to the situation at hand, such as screening or other procedures (e.g. temporary protection) and, in some circumstances, individual status determination."

15. UNHCR, *Prima Facie Guidelines*, para. 36: "A prima facie approach remains appropriate as long as the readily apparent circumstances prevailing in the country of origin or former habitual residence continue to justify a group-based approach to refugee status. The decision to adopt a prima facie approach, therefore, needs to be kept under periodic review, such that the on-going use of the practice is deliberative. Likewise, through registration, the profile of individuals and their reasons for flight can be monitored on a continual basis."

16. UNHCR, *Prima Facie Guidelines*, para. 37, "When circumstances change, careful consideration of ending the prima facie approach needs to be undertaken. Such reviews are guided by the situation in the country of origin, while recognizing the need for consistency and stability in refugee status approaches."

17. UNHCR, *Prima Facie Guidelines*, para. 17.

References

Amit, R., & Landau, L. (2016). Refugee protection is politics. Retrieved from https://www.opendemocracy.net/openglobalrights/roni-amit-loren-b-landau/refugee-protection-is-politics.

Berger, I., Hepner, T., Lawrance, B., Tague, J., & Terretta, M. (Eds.). (2015). *African Asylum at a Crossroads: Activism, Expert Testimony, and Refugee Rights*. Ohio University Press.

European Commission. (2016). Commission Recommendation of 15.6.2016 addressed to the Hellenic Republic on the specific urgent measures to be taken by Greece in view of the resumption of transfers under Regulation (EU) No. 604/2013, C(2016) 3805 final. Retrieved from http://www.refworld.org/docid/576a990d4.html.

Frelick, B. (2016). Political realities challenge refugee reform. Retrieved from https://www.opendemocracy.net/openglobalrights/bill-frelick/political-realities-challenge-refugee-reform.

Hathaway, J. (2005). *The Rights of Refugees Under International Law*. Cambridge University Press.

Hathaway, J. (2016). A global solution to a global refugee crisis. Retrieved from https://www.opendemocracy.net/openglobalrights/future-of-refugee-protection.

Hathaway, J., & Foster, M. (2014). *The Law of Refugee Status* (2nd ed.). Cambridge: Cambridge University Press.

Jackson, I. (1999). *The Refugee Concept in Group Situations*. Leiden, The Netherlands: Martinus Nijhoff.

Lawrance, B., & Ruffer, G. (Eds.). (2015). *Adjudicating Refugee and Asylum Status: The Role of Witness, Expertise, and Testimony*. Cambridge University Press.

Rusu, S. (1989). The Development of Canada's Immigration and Refugee Board Documentation Center. *International Journal of Refugee Law, 1*(3).

Rusu, S. (1994). Refugees, Information and Solutions: The Need for Informed Decision-Making. *Refugee Survey Quarterly, 13*(1).

Rusu, S. (2007). *Decision-Making in Refugee Cases Collection of Texts from the EU-Macao Cooperation Programme in the Legal Field*. Macao: Centro de Formacao Juridica e Judiciaria.

Rutinwa, B. (October 2002). *Prima Facie Status and Refugee Protection*, http://www.unhcr.org/3db9636c4.pdf.

Simeon, J. C. (Ed.). (2013). *The UNHCR and the Supervision of International Refugee Law*. Cambridge University Press.

Stainsby, R. (2009). UNHCR and Individual Status Determination. *Forced Migration Review*, April, 52–53.

UNHCR. (2014). Providing for Protection: Assisting States with the Assumption of Responsibility for Refugee Status Determination—A Preliminary Review, PDES/2014/01. Retrieved from http://www.refworld.org/docid/53a160444.html.

UNHCR. (2015). *Guidelines on International Protection No. 11: Prima Facie Recognition of Refugee Status*. Geneva, Switzerland.

UNHCR. (Reissued December 2011). *Handbook on Procedures and Criteria for Determining Refugee Status Under the 1951 Convention and the 1967 Protocol Relating to the Status of Refugees*. (HCR/1P/4/ENG/REV.3). Geneva, Switzerland.

3 Biometric Identification, Displacement, and Protection Gaps

Lindsey N. Kingston

The circumstances of forced displacement often leave individuals without state-issued documentation that can provide basic biographical information and attest to their legal status in a territory, which can exacerbate vulnerabilities and lead to devastating human rights consequences. Indeed, inadequate documentation starkly highlights a person's lack of *functioning citizenship*—an active and mutually beneficial relationship with a government, which is vital for enjoying full membership in a political community and accessing fundamental rights (Kingston, 2014). Without a duty-bearing state to advocate for their rights, refugees and other displaced persons frequently rely on the international community to fill protection gaps—including the provision of food, shelter, and basic health care—and must regularly prove their identities and legal statuses in order to access such goods. With these needs in mind, technological advancements in the field of biometrics offer new options for identification and are increasingly posited as a "solution" for documenting refugees and other vulnerable populations. Biometric systems measure physical or behavioral characteristics—including fingerprints, faces, irises, and DNA—and are often being used for identity management, either to determine a person's identity or to verify their identity claim (Jain et al., 2011; Ashbourn, 2014).[1]

Critics of biometric documentation warn of potential negative consequences, however, including the denial of humanitarian aid, limitations on coping mechanisms to survive displacement, and threats to basic human rights. In terms of refugee protection, these harmful impacts could create additional hardships for vulnerable groups who have already faced state-sponsored discrimination and violence. Some evidence suggests that the large-scale deployment of biometric identification is marked by technology failures that can lead to false identification and resulting gaps in aid provision, for instance. The use of biometrics is also complicated by legal and ethical concerns—including worries about privacy rights, "function creep," and the prioritization of security over refugee rights. Although international human rights law does not focus specifically on biometric technology, it does outline basic rights and entitlements that may inform these discussions and help formulate plans for mitigating risks. This chapter will therefore consider the role of identification in human rights protection (including relevant rights guaranteed by international law and other legal instruments) and offer future research agendas to better understand this relationship. Recommendations focus on potential risks to vulnerable populations, emphasizing the prioritization of human rights norms. Approaches for mitigating risks may include reframing biometric data as an issue of property rights (rather than solely one of individual privacy rights), while further research is needed to better understand the unintentional consequences of biometric identification on the displaced.

It is noteworthy that refugees are entitled to special protections under the 1951 Convention Relating to the Status of Refugees and its 1967 Protocol, which are invoked by millions of displaced persons each year, to at least temporarily fill the gap left by lack of functioning citizenship. A person who is recognized as a refugee under international law is entitled to a variety of critical civil and socioeconomic rights, including rights that enable them to pursue solutions to their refugeehood (Hathaway & Foster, 2014).

While recognized refugees are entitled to protections under international law, the broader category of "forcibly displaced persons" also includes other vulnerable populations denied functioning citizenship but not afforded such extensive protections under the 1951 Convention. The UNHCR monitors and often assists identified "populations of concern," including (in addition to refugees) asylum-seekers, returned refugees, internally displaced persons (IDPs) who are protected and/or assisted by UNHCR, returned IDPs, stateless persons, and various other groups in more than 180 countries. These populations are often forcibly displaced from their homes by armed conflict, discrimination and violence, and/or natural disasters. In the case of stateless persons, who lack legal nationality to any state, some stateless individuals are forcibly displaced while others remain "home" but lack legal nationality and resulting government protections (see United Nations High Commissioner for Refugees, n.d.). In many cases, protection for populations of concern is fraught with difficulty—particularly if a group faces state-sanctioned persecution within a sovereign territory. IDPs, for instance, are often displaced for the same reasons as refugees but lack international protections under the 1951 Convention because they have not crossed international borders (although IDPs do enjoy some protections under various international and regional frameworks). And although statelessness violates the "right to a nationality" outlined by Article 15(1) of the 1948 Universal Declaration of Human Rights (UDHR)—which is reinforced by a number of binding international laws, including the 1954 Convention Relating to the Status of Stateless Persons and the 1961 Convention on the Reduction of Statelessness—lack of legal nationality remains a pervasive human rights abuse that impacts at least 10 million people worldwide (United Nations High Commissioner for Refugees, 2014). In all of these cases, identity documentation is often vital for accessing basic rights and protections from state governments, as well as securing aid from humanitarian agencies.

Identity Documentation and Technology

Since the birth of the modern passport system in the early twentieth century, individuals have come to depend on state documentation to legitimize their identities—a process that often denies undocumented persons state protections and leaves them vulnerable to abuses (Torpey, 2000). Modern passports serve as "boundary objects" that standardize information and facilitate communication at securitized state borders; they have become essential for enjoying one's freedom of movement within the contemporary global mobility regime (Häkli, 2015). Biometric passports and other forms of state-issued identification are viewed as quicker, more efficient, and secure versions of previous documentation, predicated upon the need to verify identities in the face of growing security concerns (Popescu, 2015). As reliance on such documentation is both reinforced and increased, the consequences for those without access to such documentation—a tangible product of functioning citizenship—are sorely felt. Indeed, research on the forcibly displaced and the stateless highlights how identity documentation has become an essential prerequisite for the enjoyment of a variety of fundamental human rights guaranteed under international law—including rights to employment and education, health, equality before the law, freedom from discrimination, and personal security (Blitz & Lynch, 2011; Sawyer & Blitz, 2011; Institute on Statelessness and Inclusion, 2014).

The international community has been grappling with issues of documentation for a century in response to various instances of conflict and forced displacement. During the interwar period, certificates of identity (commonly referred to as "Nansen passports") were issued by the League of Nations to allow stateless refugees to travel internationally. Later termed "travel documents" in 1938, these passports represented the international community's first attempt to provide documentation that was not legitimized by any one state. Following World War II and the creation of the United Nations, provisions were made for refugee documentation under the 1951 Refugee Convention in Articles 27 and 28. In 1977, the Executive Committee of the UNHCR recommended that refugees should be informed of their status and issued with documentation certifying their refugee status as one of the basic procedures in the asylum process. Today, it is general practice for states with established refugee determination procedures to provide some form of documentation attesting to identity and refugee status, such as a refugee certificate or identity card. These documents frequently serve as evidence that an individual has the right to reside and work in a country. In states that are not parties to the 1951 Convention, however, there are rarely possibilities to obtain refugee identity documentation (United Nations High Commissioner for Refugees, 1984).

The use of biometric data is increasingly used for identity documentation and verification purposes, including among the forcibly displaced. Humanitarian organizations (as well as law enforcement and border control agencies) frequently use fingerprints, iris recognition, DNA, and facial recognition for identity management, although lesser-known techniques include voice verification, vein pattern recognition, and even keystroke dynamics that analyze the unique ways that people type on a computer keyboard (Ashbourn, 2014). Iris recognition, for instance, became widely used by the UNHCR following the fall of the Taliban regime in 2001. Tasked with helping Afghan refugees return home after living in Pakistan camps, the UNHCR eventually made iris recognition mandatory for all returning Afghans hoping to attain travel and reintegration assistance. To accomplish this, Afghan refugees' iris images were collected, digitized, and stored in the UNHCR database; in order to receive assistance, a refugee's iris would have to match their preexisting biometric file in order to prove they were entitled to humanitarian aid, as well as to ensure that no one could collect more aid than they were entitled to. Iris recognition was posited as a solution to help ensure the fair, equitable distribution of humanitarian assistance and to reach as many deserving refugees as possible (Jacobsen, 2015). In February 2015, the UNHCR announced the completed development of its new Biometric Identity Management System (BIMS), which aims to capture and store all fingerprints and iris scans from refugees and people of concern (United Nations High Commissioner for Refugees, 2015). A similar focus on the use of biometric technology has been prevalent related to refugee resettlement and national security at the state level; in the wake of deadly 2015 terrorist attacks in Paris (and ensuing debates about resettling Syrian refugees in the United States), pressure mounted to move beyond fingerprints as a sole biometric indicator and to introduce the widespread use of iris scans and DNA tests to identify refugees and other immigrants entering the U.S. (Worth, 2015).

DNA identity applications, which can be used to establish biological relationships as well as verify identity, are also being adapted for use in relation to refugee protection, migration, and humanitarian assistance. DNA testing is advanced as a tool for identifying missing persons, providing humanitarian aid, combatting illicit inter-country adoptions, preventing immigration fraud, and identifying victims of human trafficking. While standard forensic DNA analysis can take hours or days to process, "rapid DNA analysis" technologies can process profiles much more efficiently, with a turnaround time from minutes to hours (Katsanis & Kim, 2014). Despite ethical debates about its use, DNA profiling frequently factors into strategies to improve documentation of immigration applicants and refugees. Profiled DNA can be compared to claimed relatives to establish genetic relationships, compared to databases of crime evidence and profiles of wanted criminals, and uploaded into databases for future reference and identity management (Katsanis, 2013). For example, mandatory DNA testing was offered as a solution for high levels of fraud in the U.S. family reunification program. The program was halted by the Bureau of Population, Migration and Refugees (PRM) in March 2008, which had devastating consequences for African refugees resettled in the United States (Esbenshade, 2010).

Digitized birth registration also plays an important role in information and identity management among the displaced—including babies who are born in the midst of conflict. Birth registration and the right to be recognized as a person before the law are accepted under various international legal instruments, including Article 24 of the 1966 International Covenant on Civil and Political Rights and Article 7 of the 1989 Convention on the Rights of the Child. Lack of registration creates vulnerabilities to a range of human rights abuses, including the inability to access education and health care, vulnerabilities to child labor and child soldiering, early and forced marriages, and human trafficking (United Nations High Commissioner for Human Rights, 2014). Digital and mobile technologies are viewed as opportunities to streamline registration processes and improve data quality; in Uganda, for instance, a mobile phone-based registration program was recently used to raise birth registration rates. Plan International, a non-governmental organization (NGO) that advocates for universal birth registration, contends that birth registration can help record and share vital information, issuing birth certificates and establishing legal identities for children around the world. At the same time, however, it warns that digital registration creates potential threats to child protection that must be mitigated, including the possibility for identity theft and fraud, violations of privacy, persecution based on personal characteristics, exploitation and personal security violations, and various types of exclusion (Plan International, 2015). Indeed, concerns about the negative consequences of technology-enhanced identification management pervade discussions of identity documentation and verification.

Potential Unintended Consequences of Biometric Identification

Technological advancements in the field of biometrics offer new options for refugee documentation, but these technologies also come with new complexities and legal challenges. Perhaps the most basic concern is that technology failure will lead to false identification of the forcibly displaced. In the case of using iris recognition to identify Afghan refugees, for instance, critics argued that the technology's accuracy could not be guaranteed when deployed on a large scale, leading to a higher likelihood of false matches. Despite widespread use of this technology, UNHCR accounts "make no mention of this risk of falsely matched refugees nor of any measures put in in place to detect and correct for such false matches" (Jacobsen, 2015, p. 151). Similarly, a 2008 field study of a newly deployed UNHCR biometric system found that the fingerprinting system "was erratic; it worked sometimes on someone and then sometimes, even on the same person, it wouldn't work moments later"; despite these problems, the system was deemed a success by the UNHCR and used to communicate data with host governments (Hosein & Nyst, 2014, p. 16). These kinds of errors may lead to dire consequences for the forcibly displaced, who increasingly rely on biometric systems for the documentation and verification of their identities—including their refugee status, which includes their ability to access certain forms of assistance and protection.

Relatedly, biometric technologies may upend survival strategies that displaced families rely upon; sometimes, technological successes thwart social solutions to human rights problems. Although this issue has not been adequately studied in relation to biometric identification, discussions with aid workers uncover problems worthy of further exploration. For instance, adult refugees often send children or other family members to claim food rations—especially if the registered "head of household" is busy caring for younger children, recovering from an injury or illness, or helping elderly parents. With a new reliance on biometrics, however, a claimant must show up in person to provide biometric identification (such as an iris scan). This could limit coping mechanisms that refugee families have developed to share tasks and accomplish shared goals.

When biometric systems function properly, ethical concerns arise. A 2013 report for the World Commission on the Ethics of Scientific Knowledge and Technology (COMEST), a UNESCO advisory body and forum, warned that bioethics (despite the good intentions of biometric breakthroughs) held vast challenges for human rights and security:

> ... biometrics is the technology par excellence for new forms of surveillance that not only impinge on rights, privacy, expression, and well-being, but also create new government monoliths of identity administration laden, of course, with the potential for corruption and the abuse of power. Biometrics presents a series of intrinsically ethical issues; it is "intrusive" on body, and psyche of their recipients; it creates an ambiance in which everyone is potentially a criminal or suspect (thus, creating new forms of moral panic, fear, and paranoia); it impinges on dignity and often, as in the case of airport body checks, are the grounds for humiliation and embarrassment; it also creates various "states of exception" where biometrics can be instituted as a practice that does not require the consent of those it targets. (World Commission on the Ethics of Scientific Knowledge and Technology, 2013, p. 12)

"Function creep" (also known as "mission creep") refers to the tendency for a project to exceed, or creep beyond, its original purpose—often with unintended consequences. The collection of biometric data may aid authorities in identifying suspect populations and engaging in "social sorting" based on factors such as ethnicity and religion, thereby securitizing identities and reinforcing the notion of "body as information" (Ajana, 2013, pp. 5, 7). For instance, the U.S. government uses Iraqi refugee data for "homeland security" purposes rather than human security—"a practice that may not necessarily be compatible with humanitarian aims of ensuring refugee protection" (Jacobsen, 2015, p. 156). Indeed, the use of biometrics to combat terrorism following the 9/11 terrorist attacks have been a priority for the U.S. government; a 2008 report stressed the need for coordinated identification systems, citing the 9/11 Commission Report's assertion that "for terrorists, travel documents are as important as weapons" (National Science and Technology Council, 2008, p. 6). In Europe, the Eurodac project was implemented in 2003 to identify and verify the identities of asylum seekers within the European Union, using digital fingerprints to track applicants. The practice was soon extended to cover the issue of "illegal" immigration, thereby conflating the issues of asylum and immigration, and, in various ways, criminalizing asylum seekers and refugees in the process. Similar criticisms have been leveled against the UK Home Office's Applicant Registration Cards (ARCs) for asylum seekers in the United Kingdom, which require individuals to register biometric data and may serve to exclude refugees from full inclusion into British society (Ajana, 2013).

The human right to privacy is outlined in various rights frameworks, with increasing focus on privacy rights in the digital age. Article 12 of the 1948 UDHR states: "No one shall be subjected to arbitrary interference with his privacy, family, home or correspondence, nor to attacks upon his honor and reputation. Everyone has the right to the protection of the law against such interference or attacks" (United Nations, 1948). This right is further guaranteed by Article 17 of the 1966 International Covenant on Civil and Political Rights, as well as within a variety of regional treaties (such as Article 8 of the European Convention on Human Rights, which protects private and family life), state constitutions (such as the Fourth Amendment of the U.S. Constitution, which prohibits unreasonable searches and seizures), and various domestic laws. (The U.S. Privacy Act of 1974, for example, establishes a code of fair information that governs the collection, use, and dissemination of personal information in federal systems.) In December 2013, the UN General Assembly adopted resolution 68/167, "which expressed deep concern at the negative impact that surveillance and interception of communications may have on human rights," calling upon states to respect the right to privacy in digital communications (United Nations High Commissioner for Human Rights, n.d.). The UN High Commissioner for Human Rights (n.d.) has also repeatedly cautioned that state surveillance threatens individuals' rights (including rights to privacy, freedom of expression, and freedom of association) and inhibits the functioning of civil society.

Concerns regarding function creep of biometric identification raise important questions about rights to privacy in the face of refugee identity management and state surveillance. For instance, it is unclear who owns the biometric data to begin with and how it should be shared. In the case of Syrian refugees in Jordan, biometric data could theoretically belong to the government of Jordan (since the data was collected within its borders), the individual (since it is their biometric data that makes up the record), the country where the database is stored (which is often not where the data is collected, as a safeguard), the government of Syria (which has a history of reprisal against its opposition), or the countries where refugees may eventually seek asylum. Placing biometric data in the wrong hands could potentially create new vulnerabilities for refugees, while failing to share data could leave states open to security threats (Soliman, 2016). Relatedly, the potential risks associated with digitized birth registration systems highlight additional privacy concerns related to the digital mass collection and storage of personal information. Plan International (2015) warns that potential threats to child protection include identity theft or fraud, privacy violations and the loss of agency over one's personal information, and persecution based on personal characteristics such as ethnic identity.

To complicate matters further, the use of DNA testing creates new challenges for the protection of privacy. Although this technology is less commonly used to provide humanitarian assistance than are fingerprinting and iris recognition, that is likely to change in the near future—and the possibility for DNA identification raises thorny issues related to forced displacement. The collection and use of DNA from non-criminals has been criticized for encroaching on privacy rights, as well as for creating unintended negative consequences for refugees. For instance, some critics contend that databanks of DNA specimens and profile databases could be misused to target vulnerable populations or might expose hidden family relationships, such as non-paternity or misattributed parentage (Katsanis & Kim, 2014). The use of DNA to corroborate asylum seekers' accounts of their nationality has also been called into question, with critics arguing that this practice confuses legal status with biology and further entangles asylum with exclusionary immigration policies. These criticisms help explain why the UK Border Agency's Human Provenance Pilot Project (HPPP), conducted in 2009–2010, was eventually scrapped (Tutton et al., 2013). There are also cultural issues to consider; DNA testing for familial relationships, for instance, often ignore that families are social constructions that, in many cultures, are not always biological entities or are described with different vocabularies. Differences in social or political contexts may also shape a person's concerns about DNA sample submission, perhaps because they fear retribution from repressive authorities or being implicated in criminal investigations (Katsanis & Kim, 2014).

Despite these concerns, the use of biometric data has generally been deemed acceptable in state interpretations of privacy laws—although there is certainly room for debate and reinterpretation. In the United States, the Supreme Court has suggested that the collection of biometrics (including fingerprinting) does not constitute a search under the Fourth Amendment, but ambiguity remains. In *Davis v. Mississippi* (1969), the Supreme Court noted that fingerprinting was different than other searches because it didn't involve probing into one's personal life, couldn't be repeatedly collected to harass an individual, and was a more reliable crime-solving tool than other methods, such as eyewitness accounts and confessions that could be products of abuse. The Court later held that the Fourth Amendment does not extend protection to what is knowingly exposed in public, including fingerprints (Farraj, 2011). This background means that "fingerprinting of refugees and asylum seekers appears unlikely to be considered a search under the Fourth Amendment," according to Achraf Farraj (2011, p. 925). However, scientific developments may give the Supreme Court reason to reconsider that stance; for instance, biometrics may reveal some private medical information, such as certain chromosomal disorders. In *Skinner v. Ry. Labor Executives* (1989), the Court recognized that urine and blood tests could uncover medical facts and therefore intrude upon reasonable expectation of privacy (Farraj, 2011).

If the use of biometric data does indeed constitute a search under the U.S. Fourth Amendment, issues of consent and border protection are key for determining whether exceptions are legally permissible. In the case of refugees and asylum seekers, the consent exception is unlikely because consent cannot be coerced by implied threat or covert force; if a person is fleeing persecution and believes they might be removed from the United States if they do not give their consent for fingerprinting, then the process becomes inherently intimidating and the consent exception may not apply. The Supreme Court has repeatedly found that the use of fingerprinting through the US–VISIT system does not violate privacy rights, however, because searches at the border remain a right of sovereign states to examine those who cross their borders. Yet, while a search of property (such as a vehicle) is considered a routine search at the border, it's not entirely clear if fingerprinting falls into the same category—although Farraj (2011) argues that a court would be "hard-pressed to hold fingerprinting to be anything other than routine" (p. 928), since it is widely used at borders and ports of entry. The Supreme Court has declined to determine what level of suspicion, if any, is necessary for more invasive searches such as strip, body-cavity, or involuntary x-ray searches (Farraj, 2011).

In Europe, various measures have sought to safeguard personal data. The EU Data Protection Directive (95/46/EC), soon to be replaced, established high levels of protection for data, for instance, yet the public good may take priority over individual privacy in certain cases. The Directive prohibits the processing of sensitive data, such as information concerning one's racial or ethnic origins, health, or sex life. In some instances, biometric data can overlap with such information; papillary patterns in fingerprints have been linked to health issues such as nutrition and cancer, for instance, while facial recognition can reveal racial or ethnic origins. The processing of biometric information can therefore go beyond its intended purposes, entering the realm of sensitive data protected by the Directive (De Hert, 2013). Paul De Hert (2013) also argues that "taking, measuring, and processing of biometrical data may also harm a person's personal feeling or experience of dignity" (p. 392). This includes discomfort with close bodily scrutiny and observation, as well as measures that may inadvertently force someone to disclose a disability or their religion (De Hert, 2013). Despite these concerns, the Directive allows for the processing of personal data without consent if it is necessary for the public interest. It therefore gives EU member states "ample authority to collect refugees' and asylum seekers' biometric information, either at national borders or otherwise" (Farraj, 2011, p. 930). In 2013, the European Court of Justice ruled that the use of fingerprints in EU electronic passports is lawful; although the Court acknowledged that fingerprinting infringed on individual privacy, it determined that such measures were necessary for securing against the fraudulent use of passports (Court of Justice of the European Union, 2013). Directive 95/46/EC is slated to be replaced on May 25, 2018 by the General Data Protection Regulation (2016/679), a reform measure meant to address protection gaps. Adopted on April 27, 2016, the Regulation is an attempt by the European Parliament, the European Council, and the European Commission to unify and strengthen data protection within the EU. It offers protections regarding the processing of personal data and on the free movement of such information (European Commission).

Recommendations and Conclusions

Although many discussions of biometrics and forced displacement are well-intentioned and focus on the provision of humanitarian aid, it is vital to remain aware of potential risks to vulnerable populations—and that includes careful attention to the tensions between national security and human security goals. Unfortunately, current discussions of this issue tend to focus exclusively on technical solutions without adequate consideration of the social contexts they are situated within. Technical advancements have created a new environment for humanitarian relief, yet we still know little about human responses to these technologies or how biometrics factor into the bigger issues of functioning citizenship and human rights protection. Therefore, it is imperative that future research in the area of biometrics and displacement combine technological innovations with social science research and rights-centered policies.

Moving forward, the use of biometric technologies must clearly prioritize the protection of fundamental human rights—with an eye toward the problem of function creep, as discussed above. Some scholars in the emerging field of international security studies contend that risk management has become a framework for governance, leading to the production of a "biometric state" that prioritizes national security over human rights protections. This biometric state is "motivated by an obsession with technologies of risk and practices of risk management" and is "defined by the prevalence of virtual borders, reliance on biometric identifiers vis-à-vis passports, trusted traveler programs, and national ID cards, as well as the forms of social sorting that accompany these maneuvers that focus on the management of bodies in contemporary border security/management" (Muller, 2010, p. 6). This frequently leads to the "disjoining" and "thinning" of the ideal and practice of citizenship, reducing political membership to identity management and technical operations rather than emphasizing relationships to governments (Ajana, 2013, p. 158)—including state responsibilities for protecting and promoting basic human rights. This highlights how policy advisors and social scientists "need to be alert to the many ways in which new technologies—especially in the field of genetics—might be taken up"; as such, "ethical and political vigilance" informed by socio-technical anticipation is key for avoiding abuses (Tutton et al., 2013, p. 749).

This political context has serious implications for the protection of human rights, and continued analysis and monitoring is necessary to ensure that risks are minimized. The UNHCR, for instance, is in the difficult position of utilizing biometric information for humanitarian purposes in a political environment that includes the securitization of refugees and an ongoing "war on terror"; donor demands and government requests "may place UNHCR in a difficult position that risks jeopardizing its ability to enact its role as the guarantor of refugee protection" (Jacobsen, 2015, p. 159). Relatedly, some argue that existing safeguards to protect identifiable biological specimens are insufficient, despite various state policies aimed at protecting such information, and that biometric profiles could be exploited to coerce or deceive individuals (Katsanis & Kim, 2014). In other cases, the increased use of DNA relationship testing to reunite families and to facilitate legal migration means that governments are faced with complex families that are not always bound together by biology. In these situations, it is imperative that immigration policies consider the cultural backgrounds by which families were socially constructed (Katsanis & Kim, 2014). Cultural differences are also key for understanding potential concerns about biometric data collection, which may include breakdowns in communication or suspicions about the intended uses of such information.

Social divides, such as divides between religious or ethnic groups, may lead to systems that exclude certain groups, create misunderstandings or mistrust, and even lead to systematized oppression or social violence (Plan International, 2015).[2]

One potential approach for mitigating these risks is to reframe biometric data as an issue of property rights, rather than relying solely on individual protections of privacy. This argument has been occasionally raised (particularly in relation to medical research, in which biometric data could have commercial value), but it has not been fully explored in the contexts of forced displacement and refugee identification. This is complicated territory; the property rights approach could find strikingly different legal bases depending on national jurisdictions, and further research is required to better understand the feasibility of property rights enforcement (Liu, 2009). Since ownership of biometric information is often linked to ethical arguments about human dignity, property rights in this case may more appropriately be connected to control over data—including transparent protections against unauthorized collection and assurances that subjects control the use of that information (Kindt, 2013). The inherent vulnerabilities of forced displacement make such enforcement difficult and politically charged, yet the adoption of a property rights frame could serve to empower refugees and provide them with more control over their biometric data. This approach warrants further consideration from academics and policy makers.

Further research is also necessary for better understanding the unintentional consequences of biometric identification on displaced populations. For instance, biometric identification is posited as a way to eliminate fraud—such as ensuring that documented refugees are provided their allotted food aid—but it is not fully understood how fraud has been used as a survival mechanism for displaced families, or how biometric technologies may curtail methods for responding to these situations of vulnerability. As noted, iris recognition and other technologies may place undue burdens on heads of household, requiring specific individuals to claim aid and perhaps eliminating coping mechanisms necessary for covering needs such as child and elderly care. In what ways have biometric technologies changed the daily lived experiences of displaced persons? At the same time, more information is needed to understand the full scope of technology failure—including in cases related to human behavior. A classic example among aid workers relates to cooking burns; if a refugee burns their fingers while cooking, it may take several days before their fingerprints are accurately identifiable. What other possibilities for technology failure exist? And what safety nets can help ensure that aid is not denied in these cases?

Writing from a social science perspective, it is also important to note that not every humanitarian problem has a technological solution. Inherent to the issue of identification is the question of how the international community recognizes people as worthy of special human rights protections in the first place. Despite widespread acceptance of universal human rights norms, in reality it is individuals without functioning citizenship to a duty-bearing state who bear the brunt of rights abuses. The 1951 Convention is meant to fill protection gaps for border-crossing refugees who lack this sort of political membership, but clearly their vulnerabilities make them susceptible to a range of rights concerns—including potential abuses of their rights to privacy and property. These problems are further exacerbated by the growing anti-refugee sentiments in Europe and North America, due in large part to worries regarding an influx of Syrian refugees in Europe and terrorist attacks by Islamic fundamentalist groups. The use of iris recognition and other biometric technologies not only risks "rendering the digitalized, newly accessible and newly traceable body synonymous with the kind of body to be regarded as deserving of humanitarian protection," but current discussions often provide insufficient discussion of how these processes "may in turn place new demands on UNHCR in its effort to effectively protect this type of refugee body" (Jacobsen, 2015, p. 159). Tensions between refugee rights and national security complicate the use of biometric information, prompting serious questions about the universality of rights—including the right to asylum—and confronting the international community with ethical considerations that strike at the heart of human rights norms. Is it possible to fill the protection gaps left by lack of functioning citizenship, particularly in the face of global terrorism? If not, what are the ramifications for the global human rights regime that is centered on the concept of universal and inalienable entitlements?

While it is imperative to mitigate risks associated with biometric identification, it is also necessary to view these issues through the critical lens of international human rights norms. As this chapter has highlighted, concerns associated with biometric information are intricately linked to political context, including lack of functioning citizenship and ethical ambiguity over the proper use of biometric data. In particular, security perspectives focused on migration and terrorism influence government approaches to biometric identification, and state security often conflicts with rights protection and humanitarian assistance. As such, it is necessary that any discussion of biometrics—whether it is regarding scientific development, or the formulation and implementation of policy—be informed by human rights law and intentionally focused on its intended humanitarian goals. Ideally, the driving question for scientists and policy makers alike should be: "How can we use technology to protect vulnerable people and uphold their basic human rights?" Answering that question will mean confronting difficult realities about the nature of human rights protection, national security, and respect for the forcibly displaced.

Notes

1. In 2011, the United Nations High Commissioner for Refugees (UNHCR) began issuing "smart cards" to Afghan and Burmese refugees and asylum-seekers living in India. The cards included a picture of the individual, as well as a chip encrypted with biographical data. The UNHCR estimated that 18,000 displaced people would receive these cards in India, making them less likely to be arrested or face harassment from officials (United Nations High Commissioner for Refugees, 2011). In 2012, the UNHCR and the Senegalese government launched a campaign to provide similarly digitized and biometric ID cards to 19,000 refugees. These cards included fingerprint data and were touted as a way to guarantee holders the same rights as Senegalese citizens (except for voting rights), including rights of residence and freedom of movement (United Nations High Commissioner for Refugees, 2012). The UNHCR now regularly uses fingerprinting and iris recognition technologies to identify refugees and other persons of concern.

2. To operationalize some of these recommendations and put them into practice, it is useful to consider Plan International's (2015) framework for mitigating risks. Although this recent framework focuses specifically on digital birth registration, many of its recommendations are applicable to a variety of biometric technologies. The organization focuses its strategies on three key elements that influence risk: operating environment, stakeholders, and information and technology management.

References

Ajana, B. (2013). *Governing through Biometrics: The Biopolitics of Identity*. New York: Palgrave Macmillan.

Ashbourn, J. (2014). *Biometrics in the New World: The Cloud, Mobile Technology and Pervasive Identity*. New York: Springer.

Blitz, B. K., & Lynch, M. (Eds.). (2011). *Statelessness and Citizenship: A Comparative Study on the Benefits of Nationality*. Cheltenham, UK: Edward Elgar.

Court of Justice of the European Union. (2013, October 17). Including fingerprints in passports is lawful. Retrieved from http://curia.europa.eu/jcms/upload/docs/application/pdf/2013-10/cp130135en.pdf.

De Hert, P. (2013). Biometrics and the Challenge to Human Rights in Europe: Need for Regulation and Regulatory Distinctions. In P. Campisi (Ed.), *Security and Privacy in Biometrics* (pp. 369–414). London: Springer-Verlag.

Esbenshade, J. (2010). An Assessment of DNA Testing for African Refugees. Immigration Policy Center, American Immigration Council, October. Retrieved from http://immigrationpolicy.org/sites/default/files/docs/Esbenshade_-_DNA_Testing_102110.pdf.

European Commission. Reform of EU data protection rules. Retrieved from http://ec.europa.eu/justice/data-protection/reform/index_en.htm.

Farraj, A. (2011). Refugees and the Biometric Future: The Impact of Biometrics on Refugees and Asylum Seekers. *Columbia Human Rights Law Review, 42*(3), 891–941.

Häkli, J. (2015). The Border in the Pocket: The Passport as a Boundary Object. In A. L. A. Szary & F. Giraut (Eds.), *Borderities and the Politics of Contemporary Mobile Borders* (pp. 85–99). New York: Palgrave Macmillan.

Hathaway, J. C., & Foster, M. (2014). *The Law of Refugee Status* (2nd ed.). Cambridge: Cambridge University Press.

Hosein, G., & Nyst, C. (2014). Aiding Surveillance: An Exploration of How Development and Humanitarian Aid Initiatives Are Enabling Surveillance in Developing Countries. International Development Research Centre, UK Department for International Development. Retrieved from http://www.idrc.ca/EN/Documents/WP2014-1-AidingSurveillance-web-Nov21.pdf.

Institute on Statelessness and Inclusion. (2014). The World's Stateless. Retrieved from http://www.institutesi.org/worldsstateless.pdf.

Jacobsen, K. L. (2015). Experimentation in Humanitarian Locations: UNHCR and Biometric Registration of Afghan Refugees. *Security Dialogue, 46*(2), 144–164.

Jain, A. K., Ross, A. A., & Nandakumar, K. (2011). *Introduction to Biometrics*. New York: Springer.

Katsanis, S. H. (2013). Human DNA Identity Testing Policy Report. Institute for Homeland Security Solutions, May. Retrieved from https://sites.duke.edu/ihss/files/2012/03/DNA-Policy-24Jan13_psg-sk.pdf.

Katsanis, S. H., & Kim, J. (2014). DNA in Immigration and Human Trafficking. In D. Primorac & M. Schanfield (Eds.), *Forensic DNA Applications: An Interdisciplinary Perspective* (pp. 539–556). Boca Raton, FL: CRC Press, Taylor & Francis.

Kindt, E. J. (2013). *Privacy and Data Protection Issues of Biometric Applications: A Comparative Legal Analysis*. New York, NY: Springer.

Kingston, L. (2014). Statelessness as a Lack of Functioning Citizenship. *Tilburg Law Review, 19*, 127–135.

Liu, Y. (2009). Property Rights for Biometric Information—A Protection Measure? *International Journal of Private Law, 2*(3), 244–259.

Muller, B. J. (2010). *Security, Risk and the Biometric State: Governing Borders and Bodies*. New York Routledge.

National Science and Technology Council. (2008). Biometrics in Government POST–9/11: Advancing Science, Enhancing Operations. Washington, DC. Retrieved from http://www.biometrics.gov/Documents/Biometrics%20in%20Government%20Post%209-11.pdf.

Plan International. (2015). Identifying and Addressing Risks to Children in Digiti[z]ed Birth Registration Systems: A Step-by-Step Guide. Retrieved from http://www.getinthepicture.org/sites/default/files/resources/Plan%20International%20Digital%20Birth%20Registration%20risk%20assessment%20tool.pdf.

Popescu, G. (2015). Controlling Mobility: Embodying Borders. In A. L. A. Szary & F. Giraut (Eds.), *Borderities and the Politics of Contemporary Mobile Borders* (pp. 100–115). New York: Palgrave Macmillan.

Sawyer, C., & Blitz, B. K. (Eds.). (2011). *Statelessness in the European Union: Displaced, Undocumented, Unwanted*. Cambridge: Cambridge University Press.

Soliman, S. (2016). Tracking Refugees with Biometrics: More Questions than Answers. War on the Rocks, March 9. Retrieved from http://warontherocks.com/2016/03/tracking-refugees-with-biometrics-more-questions-than-answers.

Torpey, J. (2000). *The Invention of the Passport: Surveillance, Citizenship and the State*. Cambridge: Cambridge University Press.

Tutton, R., Hauskeller, C., & Sturdy, S. (2013). Suspect Technologies: Forensic Testing of Asylum Seekers at the UK Border. *Ethnic and Racial Studies, 37*(5), 738–752.

United Nations. (1948, December 10). Universal Declaration of Human Rights. Retrieved from http://www.un.org/en/universal-declaration-human-rights/.

United Nations High Commissioner for Human Rights. (n.d.). The Right to Privacy in the Digital Age. Retrieved from http://www.ohchr.org/EN/Issues/DigitalAge/Pages/DigitalAgeIndex.aspx.

United Nations High Commissioner for Human Rights. (2014). Birth Registration and the Right of Everyone to Recognition Everywhere as a Person before the Law. Retrieved from http://www.refworld.org/docid/53ff324e4.html.

United Nations High Commissioner for Refugees. (n.d.). UNHCR Statistical Online Population Database. Retrieved from http://www.unhcr.org/pages/4a013eb06.html.

United Nations High Commissioner for Refugees. (1984). Identity Documents for Refugees. EC/SCP/33, 20 July. Retrieved from http://www.unhcr.org/3ae68cce4.html.

United Nations High Commissioner for Refugees. (2010). Convention and Protocol Relating to the Status of Refugees. Retrieved from http://www.unhcr.org/3b66c2aa10.html.

United Nations High Commissioner for Refugees. (2011, August 17). UNHCR distributes pioneering smart ID cards for refugees in India. Retrieved from http://www.unhcr.org/4e4bd1506.html.

United Nations High Commissioner for Refugees. (2012, October 22). UNHCR distributes biometric ID cards to refugees in Senegal. http://www.unhcr.org/508536389.html.

United Nations High Commissioner for Refugees. (2014). Global Action Plan to End Statelessness, 2014–2024. Retrieved from http://www.unhcr.org/en-us/protection/statelessness/54621bf49/global-action-plan-end-statelessness-2014-2024.html.

United Nations High Commissioner for Refugees. (2015). Biometric Identity Management System: Enhancing Registration and Data Management. Retrieved from http://www.unhcr.org/550c304c9.pdf.

World Commission on the Ethics of Scientific Knowledge and Technology. (2013). Ethical and Societal Challenges of the Information Society: Background document distributed for the session on Ethics of the Information Society. United Nations Educational, Scientific and Cultural Organization, May 28. Retrieved from http://unesdoc.unesco.org/images/0022/002209/220998e.pdf.

Worth, K. (2015). Can Biometrics Solve the Refugee Debate? Frontline, December 2. Retrieved from http://www.pbs.org/wgbh/frontline/article/can-biometrics-solve-the-refugee-debate.

4 Information Sharing and Multi-Level Governance in Refugee Services

Carleen F. Maitland

Displacement generates a wide range of acute needs—shelter, food, clothing, health care—just to name a few. Fulfilling these needs requires complex processes, carried out by networks of humanitarian organizations, often times in challenging circumstances. As service providers, these organizations work to coordinate efforts, leveraging areas of unique expertise, with information sharing playing a crucial role. Information sharing may be required for coordinating case management, similar to any healthcare network. It may also be necessary for managing shared responsibilities in the stewarding of aid dollars, sometimes specified through subcontracts. In these networks, successful information sharing often depends on organizational processes and policies, as well as the trust and mutual adjustment gained through working closely together in the field.

Whether working together at headquarters or out in the field, humanitarian operations are often undertaken in contexts of complex organizational hierarchies. For instance, so-called local NGOs may be a part of a larger organization, with offices in several cities. Similarly, national NGOs may have headquarters in the capital, with field offices reporting into regional operations. For international organizations, these relationships can be even more complex. These vertical, *intra*-organizational relations affect data sharing throughout humanitarian relief networks. This is particularly true where centralization results in information sharing policies and procedures being defined at higher levels in the organization (Ngamassi, Zhao, Maldonado, Maitland, & Tapia, 2010; Tapia & Maitland, 2009). As a result, effective field-level coordination and information sharing requires consideration of the hierarchies within which many organizations are embedded (Rukanova, Wigand, van Stijn, & Tan, 2015).

In addition to these organizational dimensions, information sharing is also shaped by rapidly changing technologies, presenting both opportunities and challenges (Maitland, Tapia, & Ngamassi, 2009; Tallon, Ramirez, & Short, 2013). Technological advances provide new modes of acquiring data, with implications for increasing volume, as well as new methods of storing, managing, and sharing data. In networks, leveraging these benefits is complicated by the different rates at which organizations adopt technologies. Yet, on the other hand, the need to share information across the network can serve as an impetus for adopting technologies more quickly.

For both information sharing and new technology adoption, networks and multi-level governance (that is, the distribution of power and decision-making rights at various levels of an organization) often play key roles. This is particularly the case in refugee service provision, where many service providers are international organizations, whether UN agencies or international non-governmental organizations (NGOs), working together with host country governmental organizations and local NGOs. In some cases, technology firms and volunteer technical communities might also be added to the mix.

For the international organizations, their nation-spanning nature influences structures and the ways in which field offices and operations relate to regional and international headquarters. Further, these relations, processes, and procedures are also subject to changes. Cycles of organizational change, with restructuring to enable either greater centralization or decentralization, are common. The ability of field staff to weather these changes while trying to work across organizational boundaries can be challenging.

Robust theories of multi-level governance and its implications for information sharing in refugee service provision must account for the three aforementioned phenomena: (1) motivations for and changes in processes and policies for information sharing; (2) technological change that can both help and hinder information sharing; and (3) organizational changes that affect multi-level governance. A general trend that may influence organizational change is the call for a complete overhaul of the international development and humanitarian aid sectors, often referred to as "humanitarian reform" (Barnett, 2015; Barnett, 2005; Sezgin & Dijkzeul, 2016).

With these phenomena in mind, this chapter makes the case for a research agenda that explores the role of ICTs in refugee service provision through the lens of multi-level governance. As such, it tackles some of the most pressing problems at the intersection of refugee protection and support, innovation, and technological change. The chapter examines issues reflective of the three aforementioned areas, such as: (1) how data protection policies developed in headquarters diffuse downward through an organization; (2) how new technologies such as cloud computing or biometrics diffuse through multi-level organizations and networks; and (3) how flat, nimble start-ups fully enabled by new technologies can gain traction in and perhaps create a revolution among multi-level organizations. Each of these areas requires an understanding of multi-level governance.

The research agenda is grounded in insights from extant research in organization and political science, institutional economics, and information systems. Also, to demonstrate the value of this approach, I provide two mini-cases developed from field research in the West Bank and Jordan, with the former reflecting a more stable international development context and the latter a more fluid refugee context. The former is relevant as many humanitarian organizations operate in both. Hence, procedures developed for one context may simply be replicated in another. Further, the contrast highlights differences and how information sharing is implicated by the more fluid, dynamic, and sometimes chaotic displacement environment.

Theory

Multi-level governance is significant for refugee research because ameliorating the negative effects of displacement falls to diverse organizations. Many of the most established actors are complex bureaucratic organizations, with hierarchies and interorganizational relations that span distances, in many cases nations, as well as years.

Also, unlike their for-profit counterparts, most of these organizations generate revenues through fundraising to the public, and, more importantly, through grants or allocations from donors. The donors are often foreign assistance arms of national governments, such as the U.S. State Department, DFID, and SIDA, or multilateral donors such as the World Bank. Unlike public firms who face pressure from operationally distant shareholders, mainly for profits, the influence of donors on humanitarian operations is more direct and granular, frequently occurring at the project or portfolio level. This relationship creates a greater demand for accountability, which impacts both multi-level governance, information sharing,

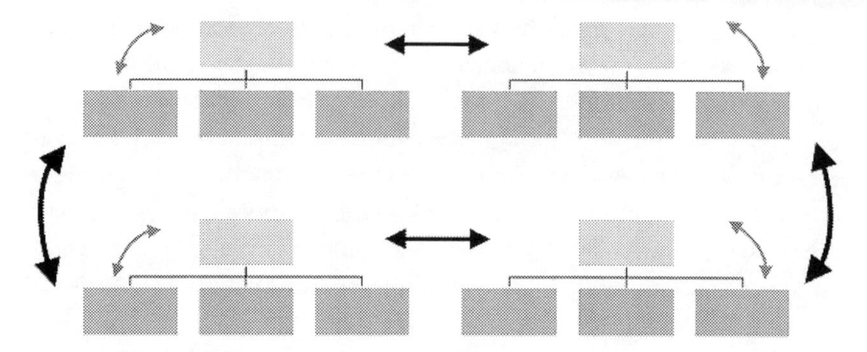

Figure 4.1
Governance across organizational hierarchies.

and technology use (Madianou, Ong, Longboan, & Cornelio, 2016; Tapia & Maitland, 2009).

In this way, multi-level governance influences hierarchical relations within organizations, as well as between donors and primary contractors and in turn subcontractors. At each subordinate level, it may also have implications for partners. Figure 4.1 depicts the implications for governance across organizational hierarchies, both intra- and interorganizational vertical relations. There are also effects for both intra- and interorganizational horizontal relations.

As employed here, multi-level governance is informed by three distinct areas of institutional theory, namely, organization science, political science, and economics. From each of the three, the focus is on extracting insights on the institutional context within the organizational milieu, as well as nation-states and economies that define these organizations' operational environments. Beyond these commonalities, the three perspectives provide unique and valuable contributions. From institutionalism in organization theory the focus is on institutions *within* organizations. From political science, perspectives on multi-level governance emphasize the relations between the institutional environment and policy formulation. Finally, from New Institutional Economics (North, 1990), insights are derived on multi-level governance within nation-states and its impact on (economic) outcomes.

Within organizations, institutions have regulative, normative, and cultural cognitive dimensions or pillars, existing at all levels (world, nation, etc.). Here the primary focus is on organizational fields, that is, a group of organizations, akin to an industry, that produce similar products and services and are subject to similar forms of industry regulation. Within the field, institutions influence governance structures that determine the exercise of power, authority, and control (Scott, 1995, p. 141).

Within the three dimensions, institutions can be observed through carriers or monikers in four categories: symbolic systems, relational systems, routines, and artifacts. Within the regulative dimension, observable components of each of these four include: rules and laws, governance and power systems, protocols and standard operating procedures, and objects complying with mandated specifications (Scott, 1995).

The organizational structure and, in turn, governance are also influenced by an organization's need for legitimacy. Each of the regulative, normative, and cultural-cognitive institutional dimensions has an associated source of legitimacy, namely, being legally sanctioned, morally governed, or culturally supported (Scott, 1995, p. 52). For humanitarian organizations, legitimacy is partly determined by accountability, both upward to donors and downward to local beneficiaries. Oftentimes, where information gathering and sharing is concerned, humanitarian organizations are more concerned with the upward form. In any case, pursuing and achieving both can be a challenge (Heyse, 2013; Stephenson, Jr. & Schnitzer, 2009; Tapia & Maitland, 2009).

Information systems research applying the institutional lens has generated four types of analyses: organizational effects on IT adoption, interactions between institutions and IT, the institutionalization of IT, and institutional discourse about IT (Nielsen, Mathiassen, & Newell, 2014). Here the emphasis is on the first three. In the domain of effects, research has found coercive, normative, and mimetic effects in the adoption of both interorganizational information systems (Teo, Wei, & Benbasat, 2003), as well as health industry electronic medical records (Sherer, Meyerhoefer, & Peng, 2016). In terms of interactions, Strong and Volkoff (2010) found role, control, and cultural misfits between an enterprise resource planning system and the organization. Finally, Hsu, Lin, and Wang (2015) examine institutional factors influencing interorganizational systems adoption across nations, including issues of legitimacy.

In contrast with organization science, institutionalism in political science places a greater emphasis on interacting authority structures *across* multiple institutional levels. In studies of political regional integration, for example in the European Union, institutional analyses of multi-level governance tackle problems such as the devolution of authority from the regional to national and local governments (Hooghe & Marks, 2001). In an organizational context, this could be equated to the changes resulting from a process of (de)centralization. In the humanitarian and displacement context, the devolution of authority may be a function of the UN's cluster system interacting with national refugee operations. (Boin, Busuioc, & Groenleer (2014) apply concepts of multi-level governance to analyze the optimal design of crisis response in the European Union using sense-making, coordination, and legitimacy as performance metrics.

The third and final institutional theory, New Institutional Economics (NIE), views institutions in broad social and economy-wide contexts. Formal and informal institutions are differentiated, with formal institutions associated with laws and informal institutions associated with the norms affecting economic exchange. NIE views organizations as embedded in an institutional framework that shapes incentives, opportunities, and costs (North, 1990). Effects are bidirectional: while organizations can affect (macro-level) institutional change, at the same time, the institutional framework shapes the adaptive efficiency of organizations (North, 1990, p. 81).

NIE also equates the existence of organizations to their control over information. The acquisition of information is costly and organizations/firms exist in part to reduce that cost, improving the efficiency of information flows. In the humanitarian context, this perspective explains the emergence of the Office of Coordination in Humanitarian Affairs (OCHA) within the UN, as well as coordination bodies and task forces within refugee operations.

Finally, NIE offers an emphasis on international comparative analyses, as well as analyses of the intersection of government with private firms and organizations. It has been employed in analyses of multi-level governance for communication policy generally (Bauer, 2014), the Internet (Van Eeten & Mueller, 2013), and, more recently, spectrum governance (Anker, 2017). In its international comparative mode, this perspective is important for explaining the significant differences found in refugee service operations across the globe and helps explain the challenges to scaling successful information sharing and technology use processes.

As the above-mentioned studies demonstrate, the emphasis on multi-level governance makes explicit the multi-level nature of institutions. Institutions not only exist at different levels, but also interact in a nested fashion. Influence between levels is bidirectional, with ideas, insights, and policies sometimes emerging at lower levels, and then being promoted by higher levels (or vice versa) to achieve scale across the organization. The highly decentralized nature of many humanitarian organizations and their operations give rise to influences in both directions. In fact, in its early days, UNHCR's Innovation Unit primarily focused on a bottom-up approach, identifying innovative practices and employees within its field operations and then scaling these ideas by disseminating information through higher levels of the organization.

Within humanitarian organizations, the mechanisms by which field-level power and authority are exercised are defined in the governance structures and arrangements, consisting of both formal and informal components. Formal components might include structured weekly meetings, with informal components reflecting managers having coffee together. While greater attention is typically paid to the formal components, in humanitarian operations, due to the rapid response, fluidity of personnel, and changing actors, informal components play an important role as well (Roberts, 2010).

Multi-level governance, particularly in international environments, highlights the tendency for legitimacy to become fragmented. For example, a humanitarian organization with a regional headquarters may develop policies that support local legitimacy in operations in one country but create conflict in another. Such effects might be observed wherein one country, lending equipment to a local implementing partner is considered a standard procedure, whereas in another it is considered a show of superiority. This tension is partly relieved through high levels of autonomy in field operations.

Applied to the domain of information flows and data analytics, the integrated multi-level governance approach developed here provides three advantages. First, it highlights the relationship between organizational policy making, knowledge of the local context, and credible policy commitments on the part of humanitarian organizations. Second, it is suitable for the temporary, flexible, and collaborative relationships in refugee services. Third, as compared to much of the IT governance literature, it makes few assumptions about direct lines of authority or control (Maldonado, Maitland, & Tapia, 2009). This matches the fragmented control over information, and increasingly information technologies, in the humanitarian community, as technologies become more associated with programs and projects rather than centrally controlled through CIOs. Confirmed in discussions with practitioners from multiple organizations, I was able to directly observe this circumstance in the USAID office in Maputo, Mozambique. There, procurement of mapping software, ArcGIS, was managed through the mission's IT staff, who then immediately handed off all responsibility for training and maintenance to program staff.

Institutions, Governance Mechanisms, Information Flows, and Data Analytics

In practice, institutional frameworks bound the options for governance mechanisms and arrangements. Mechanisms available to enhance information flows are numerous and include monetary incentives, practices and procedures, training, allocation of authority, departmentalization, specialization, and coordination teams (Foss, Husted, & Michailova, 2010). Further, a duality exists between information flows and governance, as governance mechanisms create demands for information flows and vice-versa. For example, establishment of a standard operating procedure (SOP) (governance mechanism) may in turn trigger reporting requirements (information flows) to assess compliance. Yet, at the same time, information flows/requests themselves generate demands for governance, defining who, where, why, and how information shall be provisioned.

While information flows and sharing represent particularly important organizational goals in refugee service provision, sharing in and of itself rarely affects operations. It is only when data and information are analyzed and these analyses affect decision-making that real value is generated. Both the types of analyses, and how, when, and where they are used in decision-making are often influenced by governance mechanisms, such as required reports. Absent such specifications, data may and often do lie dormant, generating inefficiencies where resources expended to collect and share data generate little value. As more data become available, the rush to collect, share, and store data may take up all available time and resources, leaving little emphasis on analyses, let alone integrating their outcomes into decision-making (Hellmann, Maitland, & Tapia, 2016; Siemen, dos Santos Rocha, van den Berg, Hellingrath, & de Albuquerque, 2017).

A hurdle to pursuing effective analyses is the necessary expertise. The field of data science is emerging to fill this gap (Meier, 2015). However, similar to the early days of IT, humanitarian organizations may face challenges in hiring and keeping necessary personnel due to the demand and salaries likely to follow. Not only is having the expertise important, but its location within the organizational hierarchy is also likely to affect when, where, and why analyses are conducted. As suggested by the following case studies, the location of this expertise is often defined by institutions and governance.

Empirical Evidence

While a detailed report of empirical research is beyond the scope of this chapter, the following cases demonstrate the value of this institutional approach, shedding light on the relationships between humanitarian organizations, multi-level governance mechanisms, and their implications for information flows and analytics. The two mini-cases were conducted in 2013 and 2015–2016 respectively, with data collected via face-to-face and Skype-based interviews, as well as field observations.[1]

Case 1: Geocoded Data Sharing among USAID Partners in the West Bank

USAID is the U.S. federal government foreign assistance organization, operating globally through its headquarters in Washington, DC and field officers, also known as "missions," affiliated with U.S. embassies. In 2012, the organization developed a new headquarters-based GeoCenter, the goal of which was promoting the use of geocoded data and geospatial analyses throughout the global organization. Within USAID, its West Bank office was considered an exceptionally avid collector and user of geocoded data. This case investigates whether and how governance mechanisms influenced this use.

The analysis examined governance mechanisms at three levels: between USAID headquarters and its West Bank field office; between the field office and its NGO partners; and between the NGO partners and their headquarters organizations.

Between USAID headquarters and the West Bank, both elements of centralized and decentralized control were observed. In the management of general forms of data, centralized processes exist in requirements for regular Performance and Planning Reports (PPRs), as well as irregular requests for additional information from Congress. However, at the same time, the field office exercised a great deal of autonomy and control over data management related to projects through its Monitoring and Evaluation (M&E) processes. In these decentralized approaches, the mission is free to define many, but not all, metrics and analyses. It was also free to develop its own information system, GeoIMS, to facilitate entry, sharing, and reporting of these data. As noted by a West Bank/Gaza staff member:

> We're highly decentralized. Also, our implementation partners—in many cases you'll find it's different for each organization. How information is used is based on decentralized relations, [and] a high level of turnover of staff. These effect how we collect information.

As for geocoded data and spatial analyses, the headquarters' GeoCenter saw its mandate as promoting spatial thinking and spatial analyses beyond mapping. They resisted the desire of the field organizations to have a centralized mapping and spatial analysis center, instead emphasizing the importance of *local* expertise and analytic capacity.

Despite having the reputation within USAID as a leader in geocoded data collection and use, the West Bank office capacity was largely limited to making maps, which were automatically generated by their management information system. Despite the general push from headquarters, the office was resistant to performing more complex spatial analyses as they felt they were unnecessary. The decentralized nature of the relationship made this stance acceptable.

As for the relationship between USAID's West Bank office and its implementing partners, depicted in figure 4.2, it is characterized by high levels of control, often specified in contractual agreements defining data flows across these organizational boundaries. Data requirements for implementing partners are defined in the Performance Management Plan (PMP), with specifications defined both by USAID headquarters and the field office. Partners are required to submit geocoded data, as stipulated in their contract. When a contract is signed, USAID's West Bank office provides training on the PMP, the management information system, and a process called the Data Quality Assessment (DQA).

The governance mechanisms embedded in the DQA are reflected in extensive processes to ensure data quality. Whereas the USAID field office conducts a DQA on its partners only once every three years, the partners

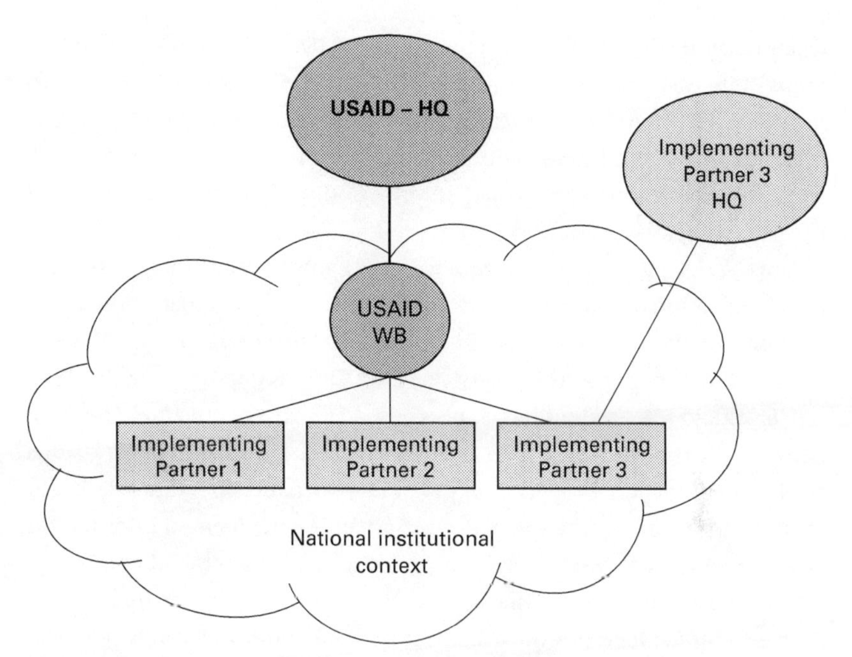

Figure 4.2
The relationship between USAID's West Bank office and its partners.

themselves conduct these analyses regularly, both internally as well as with their own subcontractors. Implementing partners conduct Data Quality Assessments internally (in one case monthly) and with subcontractors. As noted by one implementing partner:

> We have to validate information carefully. Currently we have 170 activities in GeoIMS, 60 of which are ongoing, requiring monthly inputs. We conduct our Data Quality Assessments monthly and then update the data for the activities.

As concerns the relationships of the partners with their own headquarters, the data requirements of the USAID field office are more granular than required by their headquarters. It is also important to note, partners have contracts with multiple funding agencies, only one of which is USAID. As a result, implementing partners operate multiple data management systems with multiple sets of requirements. In relation to USAID, the partners manually reconciled data between USAID's and their headquarters' information systems, undertaking the detailed and somewhat mind-numbing work of confirming consistency across cells in separate spreadsheets. Also, despite having to report geocoded data to USAID, which makes generating maps quite easy, none of three partners interviewed shared these data with headquarters, as it was neither requested nor offered. In one case, a new information system funded and designed by headquarters, once installed, would allow the entry of geocoded information, an indication it was valued by the central administration.

This case demonstrates several governance mechanisms that cross vertical organizational boundaries and their impacts on data, data flows, and analyses. In the category of organizational design, the governance mechanism of autonomy existed between USAID headquarters and its field office, and, to some extent, it was also observed between the implementing partners' field offices and their headquarters. The positive effects of this arrangement include local innovation, and systems and processes designed to fit the local context. Negative impacts include a fragmented information environment, with different systems for different purposes and extra work reconciling them all. Also, the benefits of local innovation do not always accrue to higher levels of the organization, and autonomy can hinder the adoption of centrally sanctioned innovations.

In the category of reports, USAID's headquarters' Performance Planning Reports (PPRs) and the West Bank office's Performance Management Plans (PMPs) served as governance mechanisms influencing relationships with units lower in the hierarchy. On the positive side, these reports help standardize ontologies and data structures, which can be crucial for sharing and analyses. However, when combined with autonomy, this standardization can also generate fragmentation, as local entities seek to or need to generate separate systems designed for the local context and requirements.

In the category of procedure, the USAID West Bank office Data Quality Assessment (DQA) procedure served as a governance mechanism with important effects on data quality. First, the DQA established an enforcement procedure for reports. Often, organizations take the time to specify data standards, but enforcement can sometimes be lacking. As data quality is ensured, this enhances the quality and trust in the data, which in turn benefits data flows. Organizational units are sometimes reluctant to share data across or up the hierarchy (and in many cases rightfully so) when they are uncertain of its quality. In turn, the DQA procedure benefits all levels of partners.

Case 2: Camp-Based Information Flows

This second case examines information flows in lower levels of the hierarchy: among a UNHCR camp-based field office, its NGO partners, and Syrian refugee beneficiaries in Jordan. In general, across the globe, UNHCR operates from its headquarters in Geneva, through a network of regional offices, national offices, sub-offices and field offices. While standard operating procedures exist, there is also a widespread recognition that each crisis has unique elements, and, similar to USAID, offices have a fairly high level of autonomy.

In Jordan, roughly 80% of refugees live in urban areas, with the remaining 20% housed in one of four camps: Za'atari, Marjeeb al-Fahoud, Cyber City, and al-Azraq. Za'atari camp, the focus of this case, opened in July 2012 and houses approximately 80,000 refugees. It covers an area of 5.2 km^2 (or roughly two square miles), and is operated by roughly 20 organizations. Most, if not all, have field offices in the camp and a coordinating office in the nation's capital, Amman, about 90 km away.

Within the camp, overall coordination and management is the responsibility of UNHCR, with various UN agencies, international and local NGOs, and Jordanian government organizations providing specialized services. Nearly all organizations have a physical presence in an area near the entrance known as "base camp," which serves as the local headquarters for camp operations. Further into the camp, staff working in community centers, clinics, and training facilities typically report in to these offices in base camp. Some staff members work primarily in base camp, coordinating with other camp-based organizations, and interacting with offices in Amman and beyond. These staff may have offices in both Amman and the camp. Other staff members, typically lower in the organizational hierarchy, work primarily at locations deeper in the camp and may have a small office or gathering space in base camp.

Interestingly, and with implications for information sharing, there are language divides among the staff. Those interacting primarily with refugees must speak Arabic, but not necessarily English. However, those staff members working in base camp typically speak English and Arabic, although some do not speak Arabic. With many of the highest-level meetings being held in English, speaking Arabic is not a requirement to work in base camp, as was the case with several senior administrators.

This particular case examines information flows in the process of solving problem identified by the refugee community. The refugees primarily live in caravans, divided into twelve districts, with districts further divided into blocks. The blocks are generally represented by community leaders, who are called upon to solve problems in their block or district, and liaise with humanitarian organizations when needed. Refugees share information as they go about their daily activities, and may gather information at training centers, community centers, mosques, and grocery stores, among other places.

Among the camp's humanitarian organizations, numerous governance mechanisms exist to foster information sharing. At the outset, these were almost exclusively face-to-face modes, with digital data sharing primarily occurring at offices in the nation's capital of Amman. The camp management holds a bi-weekly coordination meeting, the minutes of which are made publicly available and posted on UNHCR's headquarters' centralized information sharing portal—data.unhcr.org. During the overall coordination meeting, reports are given by sector leads, the information for which is generated by regular sector meetings. Sectors include Protection, Water, Sanitation and Hygiene (WASH), Education, and Shelter, among others. The Sector Leads may be, and often are, organizations other than UNHCR. Which organizations fill these camp-based roles is influenced by the UN-headquarters structure of the "cluster system" (IASC, 2006).

In addition to regular sectoral meetings, coordination is achieved through numerous task forces. One example is the Youth Taskforce, consisting of representatives from a wide variety of organizations providing youth services. In 2015, the Youth Taskforce held face-to-face meetings in the camp, using little, if any, information technology, although notes might be taken on a single laptop. The Taskforce's camp-based activities, primarily coordination and information sharing, involved little digital data sharing. However, they did make extensive use of email and Listservs for organizing meetings, distributing reports, and gathering feedback. More systematic or structured use, they noted, occurred in Amman, where the urban context of regular electricity and Internet access was more amenable to the medium.

Email is available in the camp, if somewhat unreliable. While electrical power has been a problem, the camp management has had access to information systems and technologies almost since the beginning. As was observed in 2014, these services were largely managed by UNHCR IT staff in the capital Amman, rather than staff located within the camp. However, over time, organizations operating in the camp brought in IT staff, and UNHCR hired its first camp-based Information Management Officer in January 2016.

Within the camp, similar to the West Bank case, organizations collect data for monitoring and evaluation (M&E) purposes. Because they have a variety of donors with differing requirements, these data are not standardized by UNHCR. In an attempt to help coordinate and share information, UNHCR established a shared database where the various organizations were to upload their M&E data. Ideally, such a database would provide a system encouraging data reuse, thereby helping reduce the amount of data collected and mitigate refugee "survey fatigue," as well as help provide a detailed perspective on operations in the camp. However, as of Spring 2016, few organizations had complied with the request to upload data into the system. The challenges of sharing M&E data in the humanitarian sector are well documented, and include competitive relationships, a disinclination to share data that might suggest outcomes were not favorable, and simply a lack of time (Maitland et al., 2009).

Despite challenges in internal data sharing and management within the camp, a plethora of information is available on UNHCR's camp-specific website, hosted on the data.unhrc.org site. The site is largely externally oriented, which has become typical in the humanitarian community. (In order to allow headquarters' staff, as well as donors, easy access to updated information, publicly available websites are used.)

For information sharing between refugees and camp staff, community gatherings are an important mechanism. These meetings are the primary forum for refugees to voice concerns, to receive updates, and to hear responses from UNHCR and implementing partners as to the status of problem resolution. Remarkably, the camp staff host 24 meetings each month, with two meetings in each of the 12 districts, one for men and one for women. These meetings are preceded by focus group meetings for each block—an enormous number of meetings, in which the problems to be addressed during the community gathering are clarified.

The invitations, minutes and reports taken at and resulting from community gatherings exemplify how multi-level governance effects information flows at lower levels of organizations. As indicated by figure 4.3, the information flows begin when notes from focus groups are transferred from

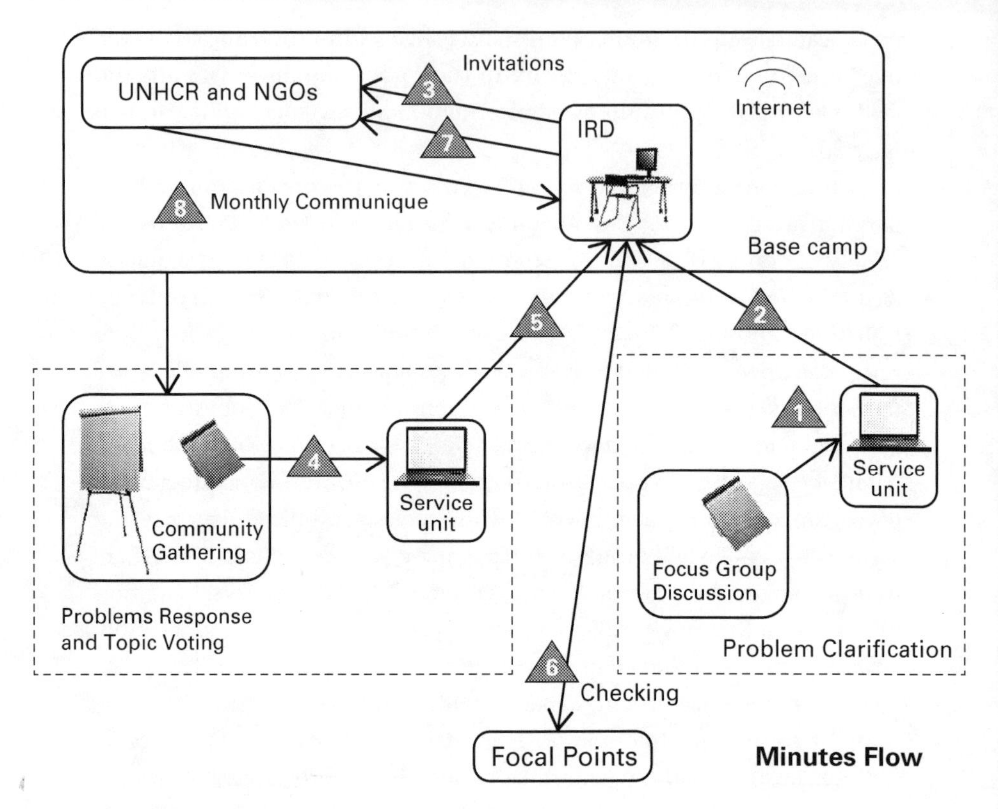

Figure 4.3
Information flows from community gatherings.

focus group discussions to district-level service units and then are passed to bosses in base camp. From there, the NGO managing community relations and a subcontractor to UNHCR, IRD, issues invitations to the meeting. The invitations are intended to spur participation from agencies that can provide explanations or status updates on problems faced by the refugees. Because UNCHR and IRD lack formal authority over all UN agencies and NGOs in the camp, response to invitations is uncertain. Following the community gathering, meeting minutes are processed through IRD, checked for accuracy by district level staff, referred to as "focal points," and then passed to UNHCR for broader distribution and integration into the monthly communiqué. Also, some of the "processing" includes transitioning notes from written to typed, combining and reconciling notes taken by several staff members, and translating from Arabic to English.

Not formally depicted in figure 4.3, and difficult to uncover, are the refugee community's own structures and governance mechanisms. Like any community, social standing influences community and government structures. There are local and religious leaders (sheiks and imams), relatively wealthy business people, sports stars, artists, and the educated. They communicate mostly face-to-face but mobile phones are ubiquitous. While the community gatherings are a means to air grievances with the camp leaders, in reality the refugees solve many problems on their own. Also, the refugee leaders have indicated that face-to-face meetings are their preferred mode of receiving information.

This case demonstrates the interactions of governance mechanisms and digital and non-digital forms of information flows, and the influence of multi-level governance. UNHCR's role in providing coordination and camp management are enshrined in UN protocols and international agreements. That role subsequently is reflected in the structure of operations in the camp. Similarly, divisions of labor within the camp are influenced by the UN cluster system structure, explaining in part the logic of networks of organizations and their relationships within the camp. Contractual mechanisms, or the lack thereof, together with the temporary nature of operations of some organizations, shape compliance and reporting behaviors.

Information flows are facilitated by four primary governance mechanisms, namely, sector-based departmentalization, task forces, coordination meetings, community gatherings, and reports. Sector-based departmentalization is reflective of structures higher in the humanitarian sector hierarchy, with norms and standards being shared, creating a greater level of continuity across crises. Task forces provide similar interorganizational sharing opportunities, but lack the higher-level structure of their sector-based counterparts. The many coordination meetings provide important mechanisms for horizontal information sharing, both between organizations and within different units of the same organization.

As compared to task forces and coordination meetings, which include only staff and therefore are influenced by higher organizational levels, the community gatherings reflect dimensions of hierarchy within the microcosm of the camp's residents. Social structures among the refugees influence who speaks at the meetings as well as the topics for discussion. That being said, in the meetings I observed there was a significant level of latitude given. Similarly, roles taken in the meetings by camp staff were reflective of their positions within their organizations. Interestingly, the gatherings also reflect the limitations of formal hierarchy. Despite the formal leadership of UNHCR in the camp, due to the lack of contractual obligations, as well as the temporary nature, they are unable to require all implementing partners attend the community gatherings. This is particularly problematic when the meeting primarily concerns a status update on a particular program, to be delivered by the responsible organization. This creates problems for information flows and is a major source of frustration for the refugees.

While this case provides evidence of the role of headquarters and international policies in the leadership structures in the camp, at the same time it reflects the high degree of local control and autonomy to establish governance mechanisms as needed. As is discussed further in chapter 7, this combination of central control and influence, together with local autonomy, is reflected in the broader context of information systems management for refugees and the displaced. For humanitarian organizations, even nominally organization-wide deployments can result in a patchwork of implementations across the globe. As an example, a 2011 audit of UNHCR's continuity of operations for its information systems notes the following:

UNHCR field offices use non-PeopleSoft applications for supporting their operations in delivering assistance to the persons of concern. These applications include systems locally developed that are considered critical to their field operations. However, an inventory of these applications was not centrally available since the Division of Information Systems and Telecommunications (DIST) does not support locally developed systems. These systems were developed locally with limited supervision from DIST for ensuring consistency and security, and had not been subject to any assessment regarding the tolerable downtime acceptable to each office in case of their unavailability. (UNHCR, 2011, p. 2)

Further, a report on Project Profile, a precursor to UNHCR's refugee registration system proGres (see chapter 7 for a detailed description) attributes differences in information systems deployments to organizational flexibility in responding to differences in operating environments.

In some field offices, where there is a large caseload and highly customi[z]ed systems are already in use (e.g., Iran and Afghanistan) the systems will not be replaced by proGres, at least in the short-term. Also for some country operations, where caseloads are small, a system like proGres, with a database server and technical support requirements, will be too advanced (and costly) to handle. (UN OIOS, 2006)

Research Directions and Conclusions

Together these two cases demonstrate the effects and limitations of multi-level governance on information sharing within and across humanitarian organizations. In addition to reflecting differences between development and displacement, the cases also highlight differences in information sharing processes for monitoring and evaluation (M&E), as in the case of USAID, versus the operational data flows managed by UNHCR. A general comparison of the development-versus-displacement contexts suggests displacement creates challenges for establishing complex information systems. Also, the M&E data management processes were extremely detailed, and, at least from the self-reports, were consistently followed. This is in contrast to data sharing related to operations, as reflected in the UNHCR managed community gathering process. There, data were not consistently managed nor analyzed. In addition, similar to USAID, UNHCR must also contend with the complex networks of partners and the resultant fragmentation of information flows. Further, the cases highlight the role of contending forces of higher control and centralization versus local autonomy and decentralization, and their sometimes complementary or conflicting results.

The cases also provide evidence of the numerous challenges large, widely distributed organizations face in making use of new and emerging information technologies. Uncertainty around how and where expertise and IT support should be provisioned, and where IT most effectively can be used, was reflected in both cases. Critical examinations of who benefits from information systems and information flows, up and down organizational hierarchies, but also between organizations and their beneficiaries, are needed.

The cases also suggest a need for greater understanding of the effects of fundamental elements of multi-level governance on interorganizational information flows. In particular, analyses of when, where, and how centralized-versus-decentralized governance are needed. Further, key questions about the changing nature of humanitarian organizations, as demands for information sharing, yet at the same time privacy and security, increase, should be investigated. Also, the role that higher-level organizational structures do and should play in fostering information sharing, through tools such as the data.unhcr.org portal, should be better understood. Issues of privacy and security in these flows are critical. Finally, given the importance of meetings to the coordination function, there is a need to investigate ways in which meetings can be better supported to help capture and share information.

Specific research questions reflecting these various domains might include:

- How do information-sharing policies developed at headquarters influence sharing in the field?
- To what extent are field operations in control of new technology adoption and how does this control (or lack thereof) influence information sharing?
- How do changes in organizational systems, particularly those promulgated by headquarters, influence field-based information sharing?
- How is network-wide refugee service provision affected by information sharing and what role does multi-level governance play? What role can coordination bodies play?
- How will trends in the global organization of humanitarian operations influence multi-level governance and in turn information flows?

As the humanitarian community responds to technological, organizational, and social change, the need, demand, and opportunities for information sharing are increasing. Understanding the potential of and limits to sharing requires a nuanced understanding of the organizations involved across all levels. Analyses conducted at only a single organizational level, focusing exclusively on headquarters, regional offices, or the field, miss vital sources of formal rules, regulations, and best practices that influence adoption and use of new technologies. Likewise, the informal mechanisms, such as informal sources of power and decision-making, must also be taken into account.

Humanitarian organizations' support of refugees within a constantly changing operational environment highlights the need to understand these dynamics. The ability of the sector to use ICTs to meet its information needs and improve refugee lives, including supporting their self-reliance, will likely result from a variety of efforts across organizational levels through various mechanisms of multi-level governance.

Note

1. As is often the case, UN, governmental, and NGO staff members involved in these interviews were hesitant to be recorded. As a result, summary notes of interviews are used in place of direct quotes.

References

Anker, P. (2017). From Spectrum Management to Spectrum Governance. *Telecommunications Policy*. doi: 10.1016/j.telpol.2017.01.010.

Barnett, Michael, & Walker, Peter. (2015). Regime Change for Humanitarian Aid. *Foreign Affairs*, July–August, 130–141.

Barnett, M. (2005). Humanitarianism Transformed. *Perspectives on Politics*, *3*(4), 723–740. doi:10.1017/S1537592705050401.

Bauer, J. M. (2014). Platforms, Systems Competition, and Innovation: Reassessing the Foundations of Communications Policy. *Telecommunications Policy*, *38*(8), 662–673. doi:10.1016/j.telpol.2014.04.008.

Boin, A., Busuioc, M., & Groenleer, M. (2014). Building European Union Capacity to Manage Transboundary Crises: Network or Lead-Agency Model? *Regulation & Governance*, *8*(4), 418–436. doi:10.1111/rego.12035.

Foss, N. J., Husted, K., & Michailova, S. (2010). Governing Knowledge Sharing in Organisations: Levels of Analysis, Mechanisms and Research Directions. *Journal of Management Studies*, *47*(3), 455–482. doi:10.1111/j.1467-6486.2009.00870.x.

Hellmann, D. E., Maitland, C. F., & Tapia, A. H. (2016). Collaborative Analytics and Brokering in Digital Humanitarian Response. In *Proceedings of the ACM 2016 Conference on Computer Supported Cooperative Work—CSCW '16* (pp. 1284–1294). http://doi.org/10.1145/2818048.2820067.

Heyse, L. (2013). Tragic Choices in Humanitarian Aid: A Framework of Organizational Determinants of NGO Decision Making. *Voluntas*, *24*(1), 68–92.

Hooghe, L., & Marks, G. (2001). *Multi-Level Governance and European Integration*. Washington, DC: Rowman & Littlefield.

Hsu, C., Lin, Y.-T., & Wang, T. (2015). A Legitimacy Challenge of a Cross-Cultural Interorganizational Information System. *European Journal of Information Systems*, *24*(3), 278–294. doi:10.1057/ejis.2014.33.

IASC. (2006). *Guidance Note on Using the Cluster Approach to Strengthen Humanitarian Response*. United Nations.

Madianou, M., Ong, J. C., Longboan, L., & Cornelio, J. S. (2016). The Appearance of Accountability: Communication Technologies and Power Asymmetries in Humanitarian Aid and Disaster Recovery. *Journal of Communication*, *66*(6), 960–981.

Maitland, C., Tapia, A. H., & Ngamassi, L.-M. (2009). Information Management and Technology Issues Addressed by Humanitarian Relief Coordination Bodies. In J. Landgren & S. Jul (Eds.), *Proceedings of the 6th International Conference on Information Systems for Crisis Response and Management (ISCRAM)* (p. 10). Gothenburg, Sweden.

Maldonado, E. A., Maitland, C. F., & Tapia, A. H. (2009). Collaborative Systems Development in Disaster Relief: The Impact of Multi-Level Governance. *Information Systems Frontiers, 12*, 9–27. doi:10.1007/s10796-009-9166-z.

Meier, P. (2015). *Digital Humanitarians: How Big Data Is Changing the Face of Humanitarian Response*. Boca Raton, FL: CRC Press.

Ngamassi, L.-M., Zhao, K., Maldonado, E., Maitland, C., & Tapia, A. H. (2010). Exploring Motives for Collaboration Within a Humanitarian Inter-Organizational Network. In *Proceedings of the 5th Annual iConference* (p. 9).

Nielsen, J., Mathiassen, L., & Newell, S. (2014). Theorization and Translation in Information Technology Institutionalization: Evidence From Danish Home Care. *Management Information Systems Quarterly, 38*(1), 165–186.

North, D. C. (1990). *Institutions, Institutional Change and Economic Performance*. Cambridge University Press.

Roberts, N. C. (2010). Spanning "Bleeding" Boundaries: Humanitarianism, NGOs, and the Civilian–Military Nexus in the Post-Cold War Era. *Public Administration Review, 70*(2), 212–222.

Rukanova, B., Wigand, R. T., van Stijn, E., & Tan, Y.-H. (2015). Understanding Transnational Information Systems with Supranational Governance: A Multi-Level Conflict Management Perspective. *Government Information Quarterly, 32*(2), 182–197.

Scott, W. R. (1995). *Institutions and Organizations. Foundations for Organizational Science*. London: A Sage Publication Series.

Sezgin, Z., & Dijkzeul, D. (Eds.). (2016). *The New Humanitarians in International Practice: Emerging Actors and Contested Principles*. New York: Routledge.

Sherer, S. A., Meyerhoefer, C. D., & Peng, L. (2016). Applying Institutional Theory to the Adoption of Electronic Health Records in the US. *Information & Management, 53*(5), 570–580.

Siemen, C., dos Santos Rocha, R., van den Berg, R. P., Hellingrath, B., & de Albuquerque, J. P. (2017). Collaboration among Humanitarian Relief Organizations and Volunteer Technical Communities: Identifying Research Opportunities and Challenges through a Systematic Literature Review. In *Proceedings of the 14th ISCRAM Conference*.

Stephenson, M., Jr., & Schnitzer, M. (2009). Exploring the Challenges and Prospects for Polycentricity in International Humanitarian Relief. *American Behavioral Scientist, 52*(6), 919–932.

Strong, D. M., & Volkoff, O. (2010). Understanding Organization–Enterprise System Fit: A Path to Theorizing the Information Technology Artifact. *Management Information Systems Quarterly, 34*(4), 731–756.

Tallon, P. P., Ramirez, R. V., & Short, J. E. (2013). The Information Artifact in IT Governance: Toward a Theory of Information Governance. *Journal of Management Information Systems*, *30*(3), 141–178.

Tapia, A., & Maitland, C. F. (2009). Wireless Devices for Humanitarian Data Collection. *Information Communication and Society*, *12*(4), 584–604. doi:10.1080/13691180902857637.

Teo, H.-H., Wei, K. K., & Benbasat, I. (2003). Predicting Intention to Adopt Interorganizational Linkages: An Institutional Perspective. *Management Information Systems Quarterly*, *27*(1), 19–49.

UN OIOS. (2006). *Project Profile*. Retrieved from http://download.cabledrum.net/wikileaks_archive/file/un-oios/OIOS-20060518-01.pdf.

UNHCR. (2011). Audit of the Arrangements for Business Continuity and Disaster Recovery for non-PeopleSoft Applications in UNHCR. Retrieved from http://usun.state.gov/sites/default/files/organization_pdf/186057.pdf.

Van Eeten, M. J. G., & Mueller, M. (2013). Where Is the Governance in Internet Governance? *New Media & Society*, *15*(5), 720–736.

5 Information Worlds of Refugees

Karen E. Fisher

I never thought that life will change dramatically in this way. I run away from death to this camp. We found ourselves in the middle of the desert. It reminds me with Palestine and Somalia, I'm afraid that we will become like them.... Sometimes I spend time with my friends talking or playing cards, when the night comes I watch TV especially news channels. When the Internet was cut, it left huge space in our life. Although our bodies are here in the camp, we live with our hearts and souls with our families in Syria.
—Za'atari Camp Diary entry, April 2016, 33-year-old man from Dara'a

On World Refugee Day 2016, the UN Refugee Agency (UNHCR) reported "Conflict and persecution caused global forced displacement to escalate sharply in 2015, reaching the highest level ever recorded and representing immense human suffering" (United Nations High Commissioner for Refugees, 2016c). Over 65.3 million people, half of whom were children, were forcibly displaced by war and persecution worldwide, marking the first time that the threshold of 60 million displaced persons has been crossed and the worst humanitarian crisis since World War II. The everyday situations comprising the Information Worlds of refugees are vast and complex. In their 2016 UNHCR report about refugees' mobile use, Vernon, Deriche, and Eisenhauer write:

The digital revolution is transforming the world but refugees are being left behind.... Having to live offline means that contact and communication with loved ones is difficult and often impossible. Without access to up-to-date information on events back in their home countries as well as in their countries of asylum, refugees cannot access basic services such as health and education or make informal decisions on how to start improving their lives. A lack of connectivity constrains the capacity of refugee communities to organize and empower themselves, cutting off the path to self-reliance. But it also constrains the kind of transformative innovation in humanitarian assistance at a time when such a transformation has never been more necessary. (p. 8)

Sophisticated methods supported by extended participatory engagement with refugees and stakeholders and triangulated data are required to understand subtleties across populations, and the implicit, hidden roles played by different people such as youth, women, elders, and mentors in refugees' social networks and their effects across wider society. Such understandings can be linked to broader economic, educational, and environmental outcomes. This chapter examines the Information Worlds of displaced people, their contexts, and the ubiquitous roles that information and communication technologies (ICTs) play in the lives of refugees. The discussion delves into factors that affect such information, including needs, creating, sharing, mis- and disinformation, using, resilience, repurposing, and empowerment, integrating diverse literature and the author's fieldwork to create a social framework for addressing a question so relevant to the world's refugee population: "What are the best ways to facilitate the best information at the right times?"

Characteristics of Displaced People

Who is displaced? In which countries? What are the passages and experiences of displaced people? What are the reasons for forced migration? The primary source for such data is the *UNHCR Annual Global Trends Report,* which integrates data from the UNHCR's own reporting, varied governments, and partners including the Internal Displacement Monitoring Center (United Nations High Commissioner for Refugees, 2016a). While forced displacement has been rising for twenty years, the UNHCR (2016c&d) pinpoints three reasons for its heavily increasing rate since 2010: (1) conflicts that cause large refugee outflows are lasting longer (e.g., Somalia and Afghanistan are in their third and fourth decades of conflict); (2) dramatic or reignited situations are occurring frequently (e.g., Syria, South Sudan, Yemen, Burundi, Ukraine, Central African Republic); and (3) solutions for refugees and internally displaced people are at fifty-year lows.

According to UNHCR Global Trends, of the 65.3 million displaced people, 3.2 million are in industrialized countries awaiting asylum decisions, 21.3 million are worldwide, and 40.8 million were forced to leave their homes but still reside in their own countries. Half the world's refugees originate from three countries: Syria (4.9 million), Afghanistan (2.7 million), and Somalia (1.1 million), totaling more than half the refugees under UNHCR's global mandate (2016a). Countries with the highest internal displacement are Columbia (6.9 million), Syria (6.6 million), and Iraq (4.4 million). The country with the biggest rise of new internal displacement is Yemen with 2.5 million people.

Of utmost significance for understanding the information landscape of forced migration is that while the world watched the unprecedented journey of over one million refugees and migrants across the Mediterranean to northern Europe, UNHCR documented that the vast majority of refugees (86%) stayed in low- to middle-income countries close to situations of conflict (United Nations High Commissioner for Refugees, 2016c). Reasons people remain in border countries are several: they hope the conflict will end and they can return home; they want to stay close to the families and properties; they cannot afford to leave, cannot legally leave, etc. Hence, Turkey, Lebanon, Jordan, and Democratic Republic of Congo hosted most refugees, especially relative to their populations. In terms of asylum requests, Germany received the most—precipitated by a federal agency tweet that it was suspending the 1990 Dublin EU convention for Syrians that they had to register for asylum in their first country of arrival (Oltermann & Kingsley, 2016), followed by the United States (mainly from people fleeing violence in Central America), Sweden, and Russia. All of these phenomena have strong implications for refugees' Information Worlds—their information needs, seeking, sharing, etc., and the role of ICTs as they transited to the EU or remained in border countries or were internally displaced.

Information Worlds

Forced displacement has no boundaries—it affects people of all ages (children, adults, the elderly), and of all abilities, races, beliefs, genders, incomes, education, and positions. However, little hard, generalizable data are available about the information needs of refugees and how they share and help one another through information and technology use. Across varied fields, from refugee studies to information science, policy and international development, public health, and education, most knowledge about how resettled refugees and displaced persons experience information focuses on their interaction with service providers in destination countries. The reasons are several: refugees are regarded as protected populations by university human subjects boards, making it difficult to obtain institutional approval for research studies; population access is challenging for reasons of mobility, culture, and language; researchers use simpler, limiting methods involving short interviews, focus groups, and questionnaires to obtain data that cannot be generalized to larger or broader populations or to delve deeper to sub-strata; few common academic domains exist for researchers to share their findings and build an integrated understanding that can inform the design of public policy, applications, and services; and, little public and private funding is available to support the research—especially with media and political portrayals swaying public opinion of migrants in a negative light. Chatty, Crivello, and Hundt (2005) in the UK, expound the theoretical and methodological challenges of working with refugee children in the Middle East and North Africa. Block, Riggs, and Haslam (2013) emphasize the intrinsic importance of involving displaced persons in as much of the research process as possible, providing displaced people with ownership of the process and their own data, and a voice in interpreting the data. Through collaborative mechanisms, researchers are better positioned to understand the nature and phenomena of displacement, and make recommendations via theory, methods, policy, and practice. However, the current forced migration crisis reveals a lack of systematic knowledge of how refugees create and share information, particularly with regard to technology, and the need for better integration of the research community itself.

So what is known about refugees' information seeking, information sharing, and use of technology? What frameworks are available for studying refugees' Information Worlds? Within the field of Information Science, the sub-field of Information Behavior is about understanding Information Worlds, recognizing the contextual factors that affect the interplay of people, place, and technology. Information Behavior addresses how people experience information in everyday contexts; it focuses on understanding the development and actualization of information needs (i.e., how information is socially created); how people seek, share, and build information; how information is managed, used and repurposed, and deemed useful in a myriad of ways. Information Behavior also addresses normative situations when people do not seek, share, or use information because of such factors as trust, repercussions, and overload. Findings from Information Behavior research are used to design information systems, policy, and services across fields.

Information Behavior provides a strong framework for understanding the Information Worlds of refugees and displaced people by integrating principles of Information Behavior (Harris & Dewdney, 1994; Case, 2012) with several key theories, notably Berrypicking (Bates, 1989), Sense-Making (c.f., Dervin, 1992), Information Poverty (Chatman, 1996), and Information Grounds (Fisher, Durrance, & Hinton, 2004). For example, according to Harris and Dewdney (1994) the following principles govern people's behavior when in need of information for everyday situations:

1. Information needs arise from the help-seeker's situation.
2. The decision to seek help or not to seek help is affected by many factors.
3. People tend to seek information that is most accessible.
4. People tend to first seek information from interpersonal sources, especially people like themselves.
5. Information seekers expect emotional support.
6. People follow habitual patterns in seeking information.

Covering a broader research-based, theoretical literature, Case (2012, pp. 375–377) contributed the following principles of information behavior that are also highly germane to understanding the Information Worlds of refugees:

1. Formal sources and rationalized searches reflect only one side of human information behavior.
2. More information is not always better.
3. Context is central to the transfer of information.
4. Sometimes information—particularly generalized packages of information—doesn't help.
5. Sometimes is it not possible to make information available or accessible.
6. Information seeking is a dynamic process.
7. Information seeking is not always about a "problem" or "problematic situation."
8. Information behavior is not always about "sense-making" either.

Collectively these principles and frameworks are useful for understanding the Information Worlds of displaced people and role of ICTs. Not only do these principles remind us to take a person- or refugee-centered, context-based approach to understanding the role of information and technology in forced displacement, they also highlight why information dissemination is not a simple transmission process. Additionally, they highlight the many cognitive, social, and affective factors at play, and the circumstance that refugees are also in need of information and technology to support fun and to counter boredom. Collectively, these principles and frameworks define one's Information World—the sum of information behavior life experiences, at any given point in time, of a single individual, or, more likely, a group sharing several commonalities. An Information World encompasses all the information needs: creating, seeking, sharing, managing, using, repurposing, avoiding, and converse behaviors, as well as associated thoughts, feelings, and corporeal/embodied and other sensations that bound a set of information incidents within a particular social, physical, and temporal setting. It is thus the milieu in which a person resides, the cultural and social norms, social types, and all the forms of information that a person creates and experiences both within him/herself and with others.

Research on Forced Displacement and Information Sharing

Research on the information sharing of refugees can be considered in three contexts: (1) during journeys, passages, and transits; (2) in camps; and (3) in asylum or destination settings. Note that these contexts are not linear or inclusive, meaning not all refugees spend time in camp settings nor do all refugees spend lengthy periods in transit using multiple modes. Moreover, not all refugee camps are the same; even those that are situated alongside each other may differ, and refugees from the same country following the same route may have very different experiences reaching the same destination.

Throughout transits and stays in camps, getting settled in new cities and towns—whether in border countries or a half-world away, refugees have the same broad everyday needs as everyone, including food, shelter and jobs, child and elder care, education, government and legal services, healthcare, utilities, communications, Internet service, meeting new people, pet care, recreation/hobbies and social activities, religion, shopping, transportation, and volunteerism. However, people in a position of forced displacement may experience wide-ranging difficulties in accessing information and carrying out these activities. Factors such as geographic location and the degree to which cognitive, affective, cultural, technical, legal, and financial barriers are amplified all interplay with complexities of a person's social support network, language abilities, education, age, gender, information, and digital literacy, etc.

Journeys and Transit Countries

Early in 2015 news media and social media reports began about refugees' journeys across the Mediterranean to Turkey, Greece, Italy, and northward through Europe. Reports such as Brunwasser (2015), Rudoren (2015), and Wendle (January 2016) emphasized the significance of smartphones, how refugees were using them to keep in touch, seek information, and leave trails for others to follow by dropping pins on WhatsApp. DaPonte (2015), for example, poignantly writes how social media helps refugees across contexts, highlighting how an Iranian changes her SIM card weekly to avoid government detection while her husband awaits decision in his UK asylum case, and how refugees rely on Viber, Skype, and WhatsApp for free messaging and perceived security protocols, especially to protect LGBT rights, find family, and self-organize.

Formal research is appearing about the experiences of the over one million refugees who migrated to Europe in 2015 and the roles of technology, connectivity, and social media. The 3M Workshop of Migrants, Marginalization, and Mobiles in Singapore (http://www.sirc.ntu.edu.sg/Pages/Home.aspx) was highly diverse in scope—researchers discussed ICTs and social science theory in conjunction with transits and destination countries in the EU, as well as situations involving security, human trafficking, low wage workers, agency, and human rights abuse in China, Korea, Hong Kong, Singapore, and Dubai. At the 2016 ACM CHI Conference in San Jose, a three-day Consortium focused on Human Computer Interaction for Development (HCI4D), with a special working group on HCI, Forced Migration, and Refugees (Fisher, Yefimova, et al., 2016a). Synergistic with the establishment of a Special Interest Group at CHI on Forced Migration and the presentation of multiple papers, these events brought together researchers from industry (such as Google and Facebook), academia, and agencies whose foci vary across borders with different populations over the past few decades, using different methodologies, perspectives, and partners. The 2017 CHI Workshop in Colorado on Forced Migration was held instead at Communities and Technology 2017 in Troyes, France (Talhouk, Fisher,

Wulf, et al. (June 2017), due to the political climate in the United States and uncertainty with VISA travel permits). The 2017 CHI HCI Across Borders Workshop (Kumar, et al., 2017) included research on ICTs and displacement in the U.S., Germany, and Malaysia.

A special issue of the journal *Social Media + Society,* "Forced Migration and Digital Connectivity in(to) Europe" (Smets & Leurs, in press), comprises papers from the perspective of LGBT refugees, use of social media to capture/report violence at borders, mapping refugees' journeys, experiences of unaccompanied minors, use of Facebook by anti-immigrant groups, comparative online identity politics, and radicalization online. In their paper, Fisher, Borkert, Yafi, and Yefimova (2017) share results from a survey with over 80 Arab refugees in a Berlin camp about their information seeking and decision-making at three points: in their home country, during their journeys/transits, and now in their host country. Research questions address the nature of migrants' temporal Information Worlds, and the role of information grounds or social spaces for obtaining needed and serendipitous information. Data are being examined both for the importance and the difficulty in finding different types of information, such as deciding when to leave, how to forge the journey, who to trust, what to bring, and destination of final immigration. Other key thematic areas include: how migrants identify mis- and dis-information; the roles of different people in acting as information and technology intermediaries (migrants themselves, kin, strangers, smugglers, and host citizens in connecting migrants in(to) Europe); and the impacts of migration on economic and entrepreneurial activity. Preliminary findings overwhelmingly emphasize: the primacy of other people, specifically knowledgeable friends, for information; smartphones, such as Samsungs, for using WhatsApp and Facebook; and the value of family and contacts for providing social support. In response to what they wish they knew if starting over, refugees' replies emphasized such issues as: "not to trust smugglers," "don't cross the water (Mediterranean), go by land," "learn German," and "take maps." Also working in Berlin, Dewitz (2016) studied the information behavior of unaccompanied minor refugees and the need and usage of smartphones before, during and after their flight to Germany. Dewitz interviewed 11 Arab males between 15 and 17 years of age at a camp in Heidelberg, using playful elements to create a youth-friendly atmosphere (e.g., youth plotted their journey on a map from their home-countries to Germany and laid down icon-cards representing

WhatsApp, Imu, Facebook, Viber, or translation apps to describe how and for what they used these apps).

Synergistic reporting occurred at the 2016 Annual Conference of the International Association for the Study of Forced Migration (IASFM; Poznan, Poland). Findings demonstrate the role of information and decision-making, in addition to important contextual factors in transit. In particular, several researchers addressed aspects of forced migration and information seeking. Belloni (2016), for example, focused on the effect of emotion on Eritreans' everyday life and transit decision-making in camps and neighborhoods in Ethopia, Sudan, and Italy. Noting a type of "collective effervescence" (Durkheim, 1912) that emerges in times when departure flows are more intense, Belloni recommends that it, along with refugees' social norms and cultural values, be used as a lens to understand migrants' trajectories and high-risk behaviors, and to frame information campaigns and crisis response. Kuschminder (2016), analyzing survey data with 1056 migrants from Afghanistan, Iran, Iraq, Pakistan, and Syria (collected in Athens and Istanbul in Spring 2015), found that decision-making in transit was influenced by a wide range of factors, such as conditions in the transit country, perceived conditions in the destination country, information access and social networks, as well as policy incentives and disincentives. Also at IASFM, Jones (2016) reported on field interviews with migrants about the role of connection men—smugglers and traffickers or non-state actors who organize the travel of young West Africans (and other migrants) through Libya and by boat into Italy, emphasizing the decision-making that leads to their use, how migrants interact with them, and where, why, and how money changes hands. Freedman (2016) explored the role of gender in forced migration. Observing that women migrants face different, and, in some cases, increased sources of insecurity during their journeys, and on arrival in Europe (including sexual violence and demands for transactional sex), Freedman proposes a gendered framework of the different insecurities faced by women, but also in their decision-making and strategies for reaching Europe. Freedman analyzes EU policy responses to the migrant crisis to examine how policies may have a specific impact on women, and asks to what extent gender has been integrated into EU policies. In November 2016, Qatar University hosted the interim meeting of the International Sociological Association Research Committee on Sociology of Migration (RC31), with heavy focus on the migration experience of refugees in Gulf countries. In May 2017, the Canadian Association for Refugee and Forced Migration Studies (CARFMS) hosted its annual conference with a broad program, including ICT-focused researchers and practitioners at the Centre for Asia-Pacific Initiatives' Migration and Mobility Program, University of Victoria, British Columbia, Canada.

Camps

Refugee camps exist in many configurations: some are administered by the UNHCR, others by government agencies or corporations, depending on location. Camps vary widely in physical size and population, services available, and rules about who may live there according to specific parameters. Some camps exist for decades, such as the Palestinian camps outside Amman in Jordan that are basically settlements comprising apartment complexes. Indeed, beyond the basic senses of time and space, Turner (2015) asserts that refugee camps be explored along three dimensions: (1) camp as a place of social dissolution and a place of new beginnings where sociality is remolded in new ways; (2) the precarity of camp life in relations to the future in this temporary space; and (3) the de-politicization of life that occurs in camps due to humanitarian government, paradoxically also producing a hyper-politicized space where nothing is taken for granted and everything is contested. In short, refugee camps make a highly specialized form of Information World, akin to the closed environments such as retirement homes and prison systems studied by Elfreda Chatman (1996) and known for elements of information poverty and life in the round(ness) where thriving is based on social connectedness.

Several researchers in computer and information science have been working in camps over the past few years. Working in Dadaab, Kenya—the world's largest refugee camp—with refugees from Eritrea and Somalia, Dahya (2016) and colleagues are studying the use of ICTs by refugee educators. Dahya's landmark report reviews how ICTs are being used to support refugee education worldwide.

Sawhney and colleagues (2009), for example, created an innovative extension of the MIT Media Lab's Computer Clubhouses for Palestinian youth in the Shu'fat Refugee Camp in the West Bank and East Jerusalem, utilizing LEGO programming tools and multimedia to facilitate their digital storytelling. The ongoing program teaches digital literacy, engages teens through creativity, and empowers them to tell their own stories. Sawhney identified areas for developing design tools to support such youth media programs. Also in the West Bank, Yerousis et al. (2015) adapted the German computer club concept for use with elementary school-age children in a Palestinian refugee camp. The computer club is a well-established tool to foster learning and integration among immigrants in Germany. The researchers found that understanding local practices and values is crucial. In a separate study within the same project, Aal et al. (2015) found that involvement in computer clubs has a positive impact on girls in terms of their vocational training and social empowerment. Within the same multinational network of computer clubs, Stickel et al. (2015) reported on digital fabrication and maker spaces with marginalized children in a Palestinian refugee camp. The authors found that the use of 3D printing has benefits in developmental and educational contexts—especially relating to self-expression—but faces many socio-technical limitations.

Fisher (Fisher, 2016; Fisher et al., 2016 a&c; Fisher et al., 2017a&b; Fisher et al, under review a&b; Yafi, Yefimova & Fisher, under review), Maitland (Maitland, Tomaszewski, Belding & Fisher, et al., 2015; Maitland & Xu, 2015), and colleagues have been carrying out fieldwork at the UNHCR Al-Zataari Camp for Syrian refugees in Jordan by the Syrian border. In an in-person survey with 174 young people ages 15–24 about mobile devices and the Internet, Maitland et al., (2015a, 2015b) reported that the majority (86%) own handsets and SIM cards, as well as multiple SIMS with 79% also borrowing from friends and family. WhatsApp was most frequently used to communicate with people in both Jordan and Syria. Young people reported that Internet, social media, and online video use had all increased since moving to Za'atari; indeed, social media use had nearly doubled. The six most frequently cited Internet Apps included Google and WhatsApp. Youth indicated they would like more access to Instant Messaging/WhatsApp, as well as news sources, and the opportunity to communicate with people via social media. The Za'atari survey supports Yafi, Nasser, and Tawileh (2015), who surveyed Syrian youth about their ICT training in Syria, and derived a framework that "generates, analy[z]es, and tests data to determine to what extent ICT educational applications are valuable tools in fostering the development of key life-skills and qualifications that can improve employability among Syrian youth, empower them to actively contribute to their own development, and have a positive impact on the community around them."

Fisher and Yafi (under review) built on Fisher, Yefimova, and Bishop's (2016b) work of refugee and immigrant youth from East Africa, Myanmar, and Latin America as ICT wayfarers to understand how Za'atari youth serve as information guides for others within the camp and when living in Syria. The majority of surveyed youth had indicated they frequently help immediate family and friends, and camp/NGO staff, specifically with online education, information search (e.g., health, legal), news, employment, and mapping. To better understand the situations in which youth help others, Fisher and Yafi used Dervin's (1992) Sense-Making theory in design workshops where youth created narrative-drawings (n=50) using LEGO Mini-Figures. The depicted situations focused primarily on providing instrumental assistance—mostly of events in Syria, with fewer representations of informational, technological, and emotional assistance. The Za'atari pilot study suggests that the instinct to help others through ICT wayfaring is universal and that youth play powerful roles, including in times of crisis and displacement. As shown in figures 5.1 and 5.2, the Syrian youth shared contextual stories of how they help people in vulnerable situations—with disabilities, orphaned, without food and water, in mourning, and in poverty—by providing instrumental, informational, emotional, and technological assistance that intrinsically reflect the mandate of the Fourth Geneva Convention. The Za'atari storytelling sessions with LEGO and drawing aids proved to be a powerful and enjoyable creative tool for equally engaging both genders, despite constraints of language, culture, time, etc.

Takieddine (2014) also used a design methodology to illustrate the need to create children's spaces in refugee camps, and created a toolkit for designing such spaces. She asked Za'atari children ages 3–13 to draw their favorite places in the camp, and proposed landscape designs intended to help refugee children heal from the trauma they had endured. Takieddine wrote that the drawings allowed children to express their ideas and communicate their stories to the researcher.

In their second series of participatory design workshops at Za'atari Camp, Fisher, Yefimova, and Yafi (2016) asked 144 young people to create paper prototypes of visionary devices—Magic Genius ICT Wayfaring Devices—for helping their community and analyzed the designs for themes using qualitative techniques. The 61 drawings and specs helped provide additional context for the community's challenges, including information problems

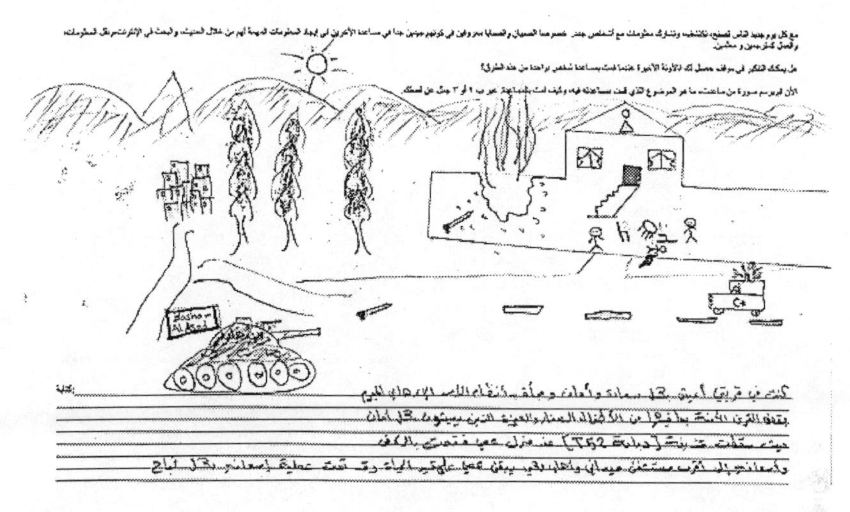

Figure 5.1
Example of narrative drawing by young Syrian male at Za'atari camp depicting situation of providing instrumental help in Syria. (Photo credit: Karen E. Fisher.)

Figure 5.2
Example of narrative drawing by young Syrian female at Za'atari camp depicting two situations: (a) helping her mother with her mobile; and (b) helping her sister with her reading. (Photo credit: Karen E. Fisher.)

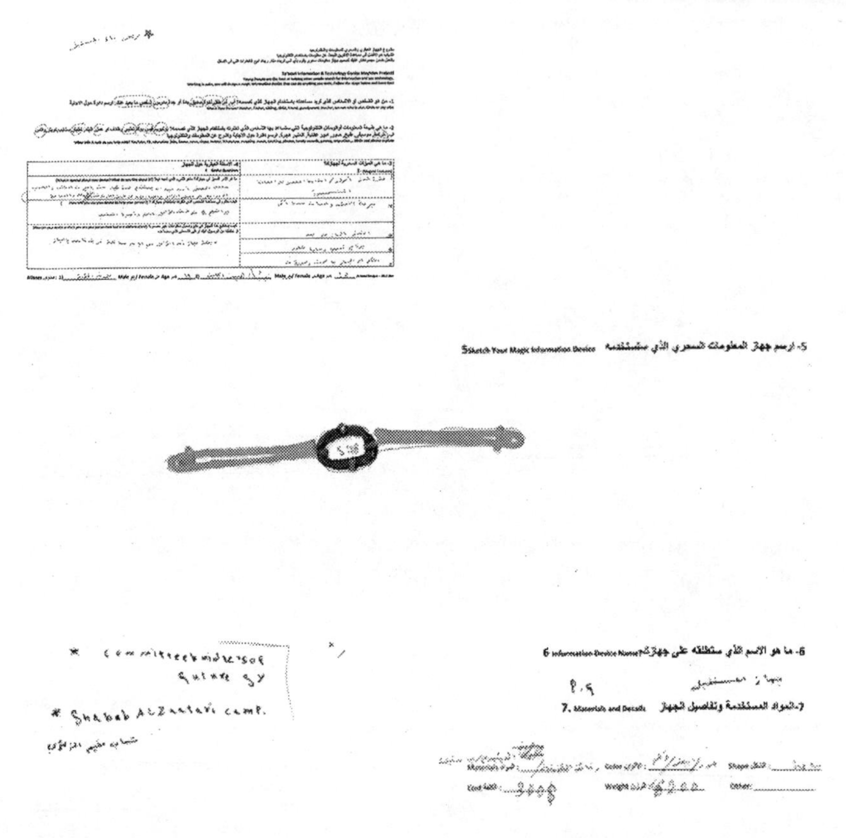

Figures 5.3 and 5.4
Example of Magic Genius spec by Syrian teens at Za'atari camp to create Magic Watch that enables social media. (Photo credit: Karen E. Fisher.)

and limited access to education. As shown in figures 5.3 and 5.4, youth created specs for glasses that detect disease; and an array of robots, watches, mobiles, magic roads, ouds, caravan heaters, and other devices that support education, social media, search, GIS, archives, future gazing, wish making, and more, while safe guarding against dis-information. Findings confirmed that Syrian refugee youth play important roles in helping others, and that support is needed to help youth with school (UNICEF 2014, 2015, UNHCR 2016b).

Most recent fieldwork at Za'atari Camp (Summer 2016) by Maitland, Fisher and colleagues (2016; Fisher et al., 2017a; under review) focuses on refugees' social capital (assets or Al Osool), their primary places for sharing information, and the best people for sharing information about particular topics. Early results indicate the overwhelming significance of people's homes for sharing information during social gatherings throughout the day, particularly in the evenings, and in markets or souks—and the perception of men as the best, most trustworthy sources of information. As Za'atari Camp is located by the Syrian border, Internet access is primarily available through WiFi at community centers. Thus, refugees express deep needs for Internet access to obtain news and communicate with others. Ongoing fieldwork by Fisher features longitudinal diaries with families and teens about their everyday lives and Information Worlds, and capacity-building, innovation technology projects based on co-design with refugees in partnership with Newcastle University Open Lab, such as the Za'atari Camp Cookbook (Fisher, 2017b; Fisher, Talhouk, Yefimova, Al-Shahrabi, Yafi, Ewald, & Comber, 2017b).

Working with people across Za'atari's 12 districts, Fisher (2016a; 2017a&b) used narrative drawings, diaries, home visits, time sheets (Day in the Life), interviews, focus groups and other methods to identify nuances in people's Information Worlds: people who have nil to small social networks, who are most isolated with scant tribal affiliations, and few interactions with neighbors are most likely to experience information poverty—as theorized by Elfreda Chatman (1996) about populations in closed settings in the US. On the other hand, people who live near or attend community centers, own shops or mobile carts, use bicycles, take classes, etc., seem to have rounder, more information rich lives. Profound differences occur in how gender affects people's Information Worlds. Boys' and men's social and physical mobility differs from girls' and women's, with males having a greater range of accessibility to places across the camp. While girls are more likely to attend school because teenage boys drop out to work (and help family income), girls are also at risk for early marriage. Figures 5.5 and 5.6 show narrative drawings of Information Worlds created by an adult male and elderly woman at Za'atari Camp. Drawings by young girls typically depict simple spaces comprising their caravans (homes), school, and community center, with maybe a mobile phone. Information Worlds of men show their caravans, friends' homes, community centers, mobile phones with apps such Facebook and WhatsApp, television, and key places throughout the camp, especially barber shops, and remote areas where cell tower strength is highest. Libraries are an emergent initiative at different NGOs across the camp driven by Syrian volunteers. These nascent libraries have strong potential to dramatically affect people's Information Worlds as library volunteers work with the community to innovate with collections, services, and outreach, including multi-generation story time that supports early literacy, home visits, mobile services, and small collections based in people's caravans to support local networks. Of key note is that book shops are absent from the Za'atari Camp landscape—books, like other materials brought into the camp, must be approved by camp security for content and are particularly screened for violence, i.e., content considered counter to the protection mandate. Similar screening/filtering occurs for digital content in computer and educational labs, much akin to schools in other countries and contexts.

Some data from Za'atari resembles Latif's (2012) findings from interviews with Palestinian refugees in Lebanese camps. Latif found that people had fonder memories of life during the war, when they were asked how relationships between Palestinians and Lebanese changed after the end of the

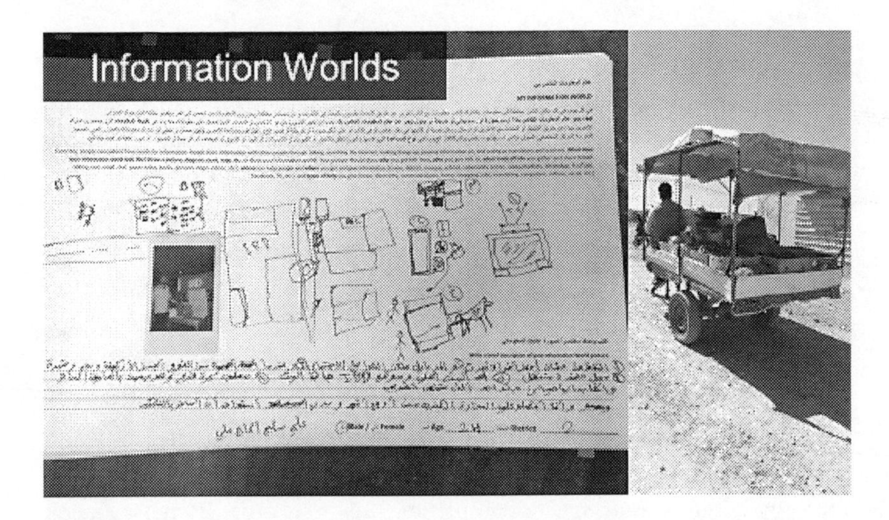

Figure 5.5
Example of Information World drawing by young Syrian man at Za'atari camp. The drawing shows a wealth of information sources, including community center, mobile coffee cart, television, friends' home, and camp perimeter where cell tower strength is strongest. (Photo credit: Karen E. Fisher.)

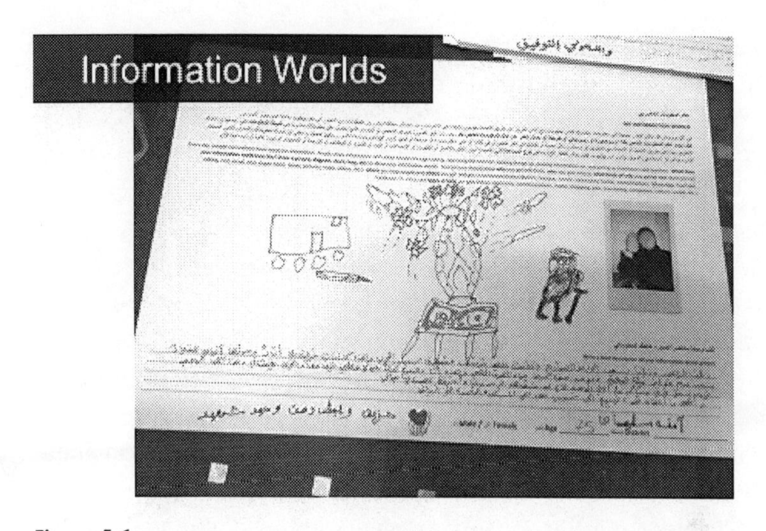

Figure 5.6
Information World drawing of an elderly Syrian woman at Za'atari camp. The drawing depicts few information sources, and she describes herself, dried flowers, her home as plain, herself as needing assistance, and with a broken, lonely heart because her son died in the Syrian conflict. (Photo credit: Karen E. Fisher.)

Lebanese civil war in 1990. At Za'atari, many older people wish to return home, and one can hear the refrain it is better to be dead in Syria than alive and alone elsewhere.

In rural Lebanon, Talhouk and colleagues (2016, 2017a) from Open Lab, Newcastle University are innovating with Syrian refugees in a small camp together with health professionals to improve antenatal healthcare using mobile devices. Based on a CitizenRadio program also deployed successfully in India by Open Lab, the aim of the Talhouk project is to use mobile technology to deliver a mobile-based radio show that features refugee(s) interviewing health experts about varied topics. The project aims to support refugee health and digital literacy while taking into account refugees' health beliefs and experiences, perceptions of negative attitudes of healthcare providers, and hierarchical and familial structures. The sensitivity of the participatory, co-design methods employed by Talhouk et al. is similar to that used in Project Lantern by Baranoff and Gonzales (2015), who also demonstrated the importance of understanding refugees' situational contexts and including refugees as co-participants in creating a cost-efficient mobile service, which combines previous generation technology with modern, near field communication capabilities that enable refugees to learn and navigate within new settings.

Destination Countries

There exists a plethora of literature on migrants' information seeking and communications across varied domains—one of the best and most comprehensive reviews is Fortunati, Pertierra, and Vincent (2011), whose monograph contains contributions about populations around the globe. In the classic work, "The Connected Migrant: An Epistemological Manifesto," Dimenescu (2008) integrated social science theory of migration to explain the transformative impact of ICTs and emergence of the connected migrant. Similarly, Srinivasan and Pyati (2007) called for studying immigrants' and refugees' diasporic information environments and reframing notions that refugees' Information Worlds are grounded in the current place-based location, but are instead, complex, globalized diaspora enabled by ICTs (thus, e-diaspora).

In one of the earliest information science studies to focus on refugees/immigrants, place, and information, Fisher, Durrance, and Hinton (2004) used Information Grounds to study the social spaces of new immigrants in Queens, New York, and found that immigrants learned important everyday Information Grounds prior to arrival through their social networks, and that trusted individuals with NGOs such as libraries were key to successful adaptation in immigrants' information behavior. Information Grounds were conceptualized by Fisher (nee Pettigrew, c.f., 1999) to explain how informal social settings facilitate the flow of information according to people, place, and information-based factors. The role of key individuals in information seeking also emerged in Fisher, Marcoux, Miller, Sanchez, and Cunningham's (2004) study of migrant Latino farmworkers in Washington State. Though the study focused on migrants, including undocumented workers, it is particularly germane to research on refugees for its use of Bates's (1989) berrypicking framework to understand migrants' reliance on new network members who have connections, knowledge and other social capital that can provide assistance. School teachers, clergy, and health professionals were identified as such important network individuals for new migrants, including those met via children's schools. According to the berrypicking framework, migrants develop a relationship, and seek/gain information and other resources, which may lead to other opportunities and situations where new contacts are made; and then the migrant's information search evolves, repeating itself, much akin to berrypicking in a patch. This notion of interpersonal berrypicking resonates with the findings of Dekker and Engbersen (2014; Dekker, Engbersen, & Faber, 2016), who examine the transformative role of social media in migrant networks. They assert that social media actively transforms the nature of migrants' social networks, thus facilitating migration in four ways: by strengthening ability to maintain strong ties with family and friends; by assisting with weak ties key to migration and integration; by establishing a new system of latent ties; and by creating rich source of discrete, unofficial, insider knowledge on migration that makes migrants "streetwise." Lingel (2015) studied migrants in New York City to understand how they became knowledgeable of urban contexts, and the connections among people, city space, and technology. She reported that wandering and getting lost ("lostness") were key to generating knowledge about urban space. Through participatory mapping and wandering of neighborhoods, as well as interviews, Lingel found that information behavior practices and use of technology shifted as individuals spatially adjusted to the city.

Focusing on the ICT wayfaring roles of refugee youth, Bishop and Fisher (c.f., 2015) held a series of multi-day community workshops, co-designing technology with immigrant and refugee teens in the U.S. and other countries. The authors developed a youth-focused interdisciplinary framework called Teen Design Days. The approach uses Design Thinking and cultural probes, aiming to highlight youth perspectives in global migration. ICTs serve as a lens for learning how teens act as information brokers in their communities—including social systems such as schools—helping their families, friends, teachers, and others navigate complex interactions. Cultural probes in the form of games, stories, sketches, and skits are incorporated in the extended design process to elicit participants' life experiences and worldviews. Similar to Talhouk et al., and Fisher et al., at Za'atari Camp, Brown and Grinter (2016) used participatory design in working with nine families from Myanmar, along with mentors and volunteer interpreters, to design Rivrtran, a "designing for transient use" messaging platform that provides "human-in-the-loop" interpretation between individuals who don't share a common language. They reported that Rivrtran was successful in mediating communication between resettling Myanmar refugee families and their host families, that scaffolding of communication via human-in-the-loop interpretation provided refugees with a short-term means of accessing diversified help outside their cultural group that mitigated the effects of cultural barriers between those communicating.

Related to Fisher et al.,'s (2016b, 2016c) work on refugee youth as ICT wayfarers, Katz (2014) focused extensively on the role of Latino youth in negotiating interactions among their families and community providers upon migration to the United States, highlighting how children's unique brokering abilities are crucial to assisting short-term family needs; but she also showed how the same responsibilities can constrain children's access to educational resources and affect their long-term goals. Working in human development, Daiute's (2010) comprehensive research with 108 war-affected youth from the former Yugoslavia reveals how workshops involving youth perspectives on society, critical narratives, and youth-designed surveys can constructively explore themes of peer and adult conflicts, and produce hypothetical community narratives. Quirke's (2012) study of information practices, through interviews with seven Afghan immigrant and refugee teens who had been in Canada for about 10 years highlights the importance of other people as primary information sources. With the viral messaging of Canada's acceptance of refugees in 2015 to refugees worldwide, it is expected that considerable fieldwork will be undertaken in Canadian contexts. Indeed, the 10th Annual Conference of the Canadian Association for Refugee and Forced Migration Studies (CARFMS), hosted by the University of Victoria in May 2017, is examining a wide range of problems relating to forced migration, including ICTs.

Extensive research has also been carried out in Australia. In Melbourne, Australia, Gifford and Wilding (2013) focused on how Karen Burmese teenage refugees use ICTs in the Home Lands digital media project. Analyzing films and photographs, they describe articulations of belonging (settlement escapes) that demonstrate how ICTs dually open up ways for becoming at home in a new country while being part of global, de-territorialized world. Earlier, Sampson and Gifford (2010) helped recently arrived refugee youth create photo-novellas and neighborhood drawings in order to understand how they connect to new places in Melbourne and that serve to aid with their resettlement. Also working in Australia with a small group of refugees, Lloyd and colleagues (2013, 2015, 2016), writing in Library Science (and stressing information literacy practices) emphasized the importance of service providers as professional mediators: "social inclusion becomes possible where information is provided via sharing through trusted mediators who assist with navigating the information landscape and information mapping, and through visual and social sources" as well as the role of place for youth. Using the same small study, Lloyd (2015) proposed the construct of "information resilience"—as an outcome of refugees' information literacy practice upon constructing new information landscapes, after transitioning over disruptions prompted by immigration—integrating research in information science and cognate fields on place, social capital, and community resilience. However, the accounts of unaccompanied minors traveling from Afghanistan and Pakistan to Australia are stories rife with information poverty and needed interventions of technology and other sources of help (Nardone & Correa-Velez, 2015). In Australia, Almohamed and Vyas (2016) interviewed five Arab refugees and two activists, identifying three common themes relevant to envisioning a supportive technology platform: (1) social isolation; (2) cultural backdrops; and (3) the role of technology, with WhatsApp and social media most prevalent. Using Sen's Capability Approach, Andrade and Doolin (2016) focused on the how ICTs promote social inclusion based on interviews with 50 refugees in New Zealand, most from Myanmar and East Africa. They devised a five-point capabilities framework that enable refugees to exercise their agency and enhance their well-being: to participate in an information society, to communicate effectively, to understand a new society, to be socially connected, and to express a cultural identity. Koo, Cho, and Gross (2011) examined how information seeking affected severe traumatic stress among North Korean refugees living in South Korea. In Malaysia, Cravero (2015) studied the use of mobile technology in support of refugee resilience, addressing how UNHCR Malaysia and other refugee assistance stakeholders could integrate mobile ICT innovation into their standard operating procedures for improved service delivery. Shankar et al., (2016) conducted a case study of a university student from Africa to understand refugee resettlement using arts-based and interview methods with the aim of improving university services and experience.

Leurs (2016) nicely summarized early refugee diaspora/border research to frame how young connected migrants and non-normative European family life are intertwined with affective human right claims of young e-diaspora. Drawing on examples from Moroccan-Dutch, Somali, and other youth, Leurs premises that young connected migrants' cross-border practices show they "do family" in a way that does not align with the typical European, and that research needs to integrate such fields as migrant studies, feminist and postcolonial theory, and digital cultures. In Germany, which accepted the most refugees of all EU countries in 2016, Neuenhaus and Aly (2017) carried out "Empathy Up (EMP UP)." This geo-location based mobile game aimed at building empathy among young Germans toward Syrian refugees, as research showed Syrians did not contact locals and vice-versa. The game addressed cultural differences within scenarios that connected players emotionally, thus minimizing prejudices by locals, increasing their willingness to get to know refugees, and making the first contact with a refugee simpler and more positive.

Value of a Smartphone

For people displaced by war and persecution, for migrants, information and ICTs are vital lifelines to the past, present, and future. According to Vernon et al. (2016) for UNHCR, most urban refugees live in places that have 2G or 3G mobile coverage; however, 20% of refugees live in areas with no connectivity. Connectivity is expensive—consuming up to a third of refugees' disposable income. Globally, refugees are 50% less likely than the general population to have an Internet-enabled phone, and 29% of refugee households have no phone at all.

Thus, accessing information is not simple, nor is accessing and finding people, especially if they live near conflict zones. Misinformation and disinformation abound as refugees and people caught in conflict fear reprisals from warring factions and from inaccurate information being shared with the public. Everything has a price. Field interviews by Rhode, Aal, Misaki, Randall, Weibert, and Wulf (2016) with 17 Syrian refugees, FSA fighters, and oppositional activists reveal the complexity and dangers of trying to communicate around the Syrian border. The authors write "Assad himself made Internet infrastructure widely available.... Today Syria seems to be divided in government-controlled parts with fairly intact infrastructures and rebel-controlled parts without terrestrial or cabled telephone or Internet access." While rebels and refugees are well-equipped with smart phones, they are unable to communicate with family if they live in rebel-controlled regions—the same finding as Fisher, Maitland, and colleagues (c. 2015 onwards) working at Za'atari Camp where refugees can pay the price of being deported if caught communicating with outsiders about verboten topics or activities.

What is a smartphone worth? In Borkert, Fisher, and Yafi's (in press) research with Arab refugees at the camp in Berlin, all respondents replied their smartphone was their most important item to have during transit.

Smartphones and free apps such as WhatsApps, Viber, and Facebook enable people to keep in touch and share information with each other about routes and borders, safe places, news, and getting settled. At the Za'atari refugee camp in Jordan, Fisher learned that mobiles are valued equally for their affordance as photo and video albums of people and times past. Mothers carried images of their sons' death photos (sharing posted death images is a common way for people caught up in civil war to identify loved ones); fathers played videos of celebrations in their homes in Aleppo. Indeed, a Magic Genius ICT Device design proposed by Za'atari youth was to create a means of storing family-community memories. Smartphones as cameras are also important for documenting and broadcasting events—such as incidents and injustices that occur at border crossings and inside borders. Rohde et al. (2016) found that:

> Low tech videos of war scenes and related atrocities play a central role of framing narratives and developments in the [Syrian] civil war. One feature of these videos is that their provenance is almost always unclear. In many if not most cases, it is hard to establish who shot the videos, who exactly is acting, or, in case they display atrocities, which side is responsible for them. Being widely disseminated via mobile devices as well as via social media platforms such as YouTube or Facebook, they are serving multiple purposes: for documentation of atrocities and war crimes, for disinformation and propaganda, for humiliation and demoralization of the enemy.

Fisher (2016) and others who carry out fieldwork with refugees report similar stories of people carrying images and video on mobiles that pertain to conflict and are shared on different bases, if at all, due to repercussions if identified. Refugees have private networks for sharing traceable images of such things as destroyed towns, graves, and acts of violence. Refugees are also highly sensitive about capturing and posting images of faces with location markers as it is impossible to predict the outcome of conflict and repercussions for people, even many years out on extended family and friends.

Conclusion: A Call for Systematic Research and Recommendations for Its Conduct

While the above synthesis reflects the growing body of scholarship on the information worlds of refugees, there is still much to be done. Our understanding and theoretical models of refugee information behavior continue to evolve as new technologies, new social norms, and cultural factors influence how information is created, sought, gathered, shared, and managed.

Critical needs for future research—in the context of Refugees' Journeys and Transits, Camps, Asylum, and Destination Settings, and the ubiquitous Value of a Smartphone—should specifically engage triangulated methods, theory, and investigators to address the multifaceted roles of people, place, time, and gender in information flow. Deeper connections are needed, linking existing studies; future work needs to be designed in the context of earlier studies and building on aggregate findings and theory, across fields that yield to useful outcomes. Throughout this chapter, research has been shared that highlights the importance of understanding the contexts of refugees' situations from their perspective, for understanding places such as Information Grounds, the roles of social types in berrypicking and social capital, and the myriad other factors affecting how refugees experience information. Based on iterative field engagement, Fisher (2016) derived topline design and field insights for working with refugees in camps by conflict zones that may more broadly apply to working with refugees in other contexts and creating participatory innovations. These insights and recommendations for factors to be considered in future research include:

1. A focus on humanitarian research—recognizing and assisting people's urgent, real needs for tangible help; sensitivities include facilitating resource drives and deliveries, providing reciprocity appropriate to the setting.
2. The disruption of social fabric—understanding how refugees' social networks, roles, and behaviors have changed due to conflict (e.g., fewer men are present in refugee camps because of home country conflict, older people cannot make the journey, larger numbers of children, increased workloads for women who may not have male relations to carry out particular tasks or engage in male-gendered spaces).
3. Closed, low-resource environments—some camps have high levels of security where Internet access and radio is not permitted, weak mobile-network access, and people have low levels of income/employment and other resources.
4. Building capacity for all—as part of working with UNHCR and refugees, the onus is to help everybody of all age, ability, race and gender.
5. The differing ubiquity of time, place, and gender—every place has its own sense of time, place, and gender based on the culture of its people and geography. In some Middle East countries, for example, weekends occur over Friday–Saturday, and the rhythm in a camp differs greatly from that in rural–urban settings, especially for its places and role of gender.

6. Iterative long-term, social engagement—working with refugees and understanding their information worlds requires investment in human relationships with multiple stakeholders and going onsite; one-shot studies with a small population provide little depth.
7. Universal design archetypes—people are the same everywhere, across time—what differs is the contexts and situations: e.g., kids everywhere design robots, but their designs vary greatly based on differing youth experiences.
8. Youth as ICT wayfarers—young people everywhere serve as information guides, helping others through technology; this is especially valid in refugee populations, and youth education and creativity needs to be promoted.
9. Connected learning—indicators of the most successful learning and community resilience occur where people feel safe and welcome in coming together, and where mentors regularly gather to facilitate interaction.
10. Innovating futures—working with refugees is not for academic exercise but about changing lives, creating futures through helping leverage community organized initiatives, livelihoods, education, and about committing to individuals.

By using these field and design insights in cognizance of the roles of individuals, Fisher asserts that social innovations and technologies can have tangible impacts on refugees' lives, instead of simply amplifying problems (Toyama, 2015). Indeed, at the outset of this chapter, the question was posed: "What are the best ways to deliver the best information at the right times?" Johnston (2016a; 2016b) provides prime research examples of how institutions are assisting refugees with information sharing by facilitating place and social interaction. Modeled on Fisher's Information Grounds, Putnam's social capital theory, and Contact Theory, Johnston explains how public libraries in Iceland, Denmark, Norway, and Sweden are hosting conversation-based programming such as Women's Story Circle, Expat Dinners, Memory Groups, and the Sprakhornan Programme (language café). Study results indicate that refugee (and library) participants benefit from unique opportunities to learn and practice language acquisition, informal information exchange, social interaction, expansion of social networks, and increase in social capital. Other examples of participatory design involving refugees, technologists, enterprise (e.g., Google), NGOs (e.g., Mercy Corps, NetHope), and municipalities are featured regularly on http://innovation.unhcr.org, with topics ranging from shelters to microfinance and livelihoods to education and human rights, among others—all shared broadly on social media. Conrad et al. (2016) describe how European universities such as TU Wien (Vienna University of Technology) responded to the refugee crisis by welcoming refugee youth, especially unaccompanied minors, to computer science classes as it enabled the youth to gain knowledge and make contact with local youth. The authors emphasize the need for flexibility, involving local NGOs who support youth, and invaluable benefits to youth, host countries, and universities. Also of core interest is Mesmar, Talhouk, Akik, and colleague's (2016) comprehensive review from refugee, provider, designer, and policy perspectives of the impact of ICTs in healthcare—focusing on innovations and gaps in humanitarian crises.

Much systematic research is needed to deeply understand the information worlds of refugees. In 2010, Caidi, Allard, and Quirke broadly reviewed research on the information practices of immigrants and refugees in the United States and Canada, characterizing the range of immigrants' information needs, methods and sources of seeking information, and the significance of interpersonal relationships, social settings, and media. They emphasize the dearth of research and the need for building systematic research programs. While the review presented here shows a nascent international research community, evident from conferences and journal special issues, much more work is needed, especially to create refugee-centered, participatory designs and services that can benefit all.

References

Aal, K., von Rekowski, T., Yerousis, G., Wulf, V., & Weibert, A. (2015). Bridging (Gender-related) Barriers: A Comparative Study of Intercultural Computer Clubs. In *Proceedings of the Third Conference on GenderIT* (pp. 17–23). New York: ACM. http://doi.org/10.1145/2807565.2807708.

Aal, K., Yerousis, G., Schubert, K., Hornung, D., Stickel, O., & Wulf, V. (2014). Come_IN@Palestine: Adapting a German Computer Club Concept to a Palestinian Refugee Camp. In *Proceedings of the Fifth ACM International Conference on Collaboration Across Boundaries: Culture, Distance & Technology* (pp. 111–120).

Almohamed, A., & Vyas, D. (2016). Designing for the Marginalized: A Step Towards Understanding the Lives of Refugees and Asylum Seekers. In *DIS 2016 Companion, June 4–8, 2016*. Brisbane, Australia.

Andrade, A. D., & Doolin, B. (2016). Information and Communication Technology and the Social Inclusion of Refugees. *Management Information Systems Quarterly, 40*(2), 405–416.

Baranoff, J., & Gonzales, R. I. (2015). Lantern: Empowering Refugees through Community-generated Guidance Using Near Field Communication. In *CHI'15 Extended Abstracts* (pp. 7–12).

Bates, M. J. (1989). The Design of Browsing and Berrypicking Techniques for the Online Search Interface. *Online Review, 13*(5), 407–424.

Belloni, M. (2016, July). Collective Effervescence among Forced Migrants in Transit: A Multi-Dimensional Analysis of Eritreans' Mobility-Related Decision-Making. In *16th Conference of the International Association for the Study of Forced Migration*. Poznań, Poland.

Bishop, A. P., & Fisher, K. E. (2015). Using ICT Design to Learn about Immigrant Teens from Myanmar. In *Proceedings of the Seventh International Conference on Information and Communication Technologies and Development* (pp. 56:1–56:4). New York: ACM. http://doi.org/10.1145/2737856.2737903.

Block, K., Riggs, E., & Haslam, N. (Eds.). (2013). *Values and Vulnerabilities: The Ethics of Research with Refugees and Asylum Seekers*. Toowong, Australia: Australian Academic Press.

Brown, D., & Grinter, R. E. (2016). Designing for Transient Use: A Human-in-the-Loop Translation Platform for Refugees. In *CHI'16: Proceedings of the 2016 CHI Conference on Human Factors in Computing Systems*. San Jose, CA.

Brunwasser, M. (2015, August 26). A 21st-Century Migrant's Essentials: Food, Shelter, Smartphone. *New York Times*. Retrieved from https://www.nytimes.com/2015/08/26/world/europe/a-21st-century-migrants-checklist-water-shelter-smartphone.html.

Caidi, N., Allard, D., & Quirke, L. (2010). The Information Practices of Immigrants. [ARIST]. *Annual Review of Information Science & Technology, 44*, 493–531.

Case, D. O. (2012). *Looking for Information: A Survey of Research on Information Seeking, Needs, and Behavior*. Bingley, UK: Emerald.

Chatman, E. A. (1996). The Impoverished Life-World of Outsiders. *Journal of the American Society for Information Science, 47*(3), 193–206. doi:10.1002/(SICI)1097-4571(199603)47:3<193:AID-ASI3>3.0.CO;2-T.

Chatty, D., Crivello, G., & Hundt, G. L. (2005). Theoretical and Methodological Challenges of Studying Refugee Children in the Middle East and North Africa: Young Palestinian, Afghan and Sahrawi. *Journal of Refugee Studies, 18*(4), 387–409.

Chib, A. (2017). (Ed.). 3M Workshop of Migrants, Marginalization and Mobiles. Singapore Internet Research Centre, Nanyang Technological University, Singapore. http://www.sirc.ntu.edu.sg/Research/3MWorkshop/Pages/Home.aspx

Conrad, K., Musliu, N., Pichler, R., & Werthner, H. (2016). Universities and Computer Science in the European Crisis of Refugees. *Communications of the ACM, 59*(10), 31–33.

Cravero, C. E. (2015). Mobile Technology for Refugee Resilience in Urban and Peri-Urban Malaysia. In *Proceedings of the Seventh International Conference on Information and Communication Technologies and Development* (pp. 34:1–34:4). New York: ACM.

Dahya, N. (2016). *Education in Conflict & Crisis: How Can Technology Make a Difference? A Landscape Review. Published by GIZ, USAID.* Germany: WVI (World Vision International).

DaPonte, J. (2015). Clear Connections: How Social Media's Power is Being Harnessed by Refugees. *Index on Censorship, 44*(1), 26–30.

Daiute, C. (2010). Critical Narrating by Adolescents Growing Up in War: Case Study Across the Former Yugoslavia. In K. C. McLean & M. Pasupathi (Eds.), *Narrative Development in Adolescence: Creating the Storied Self.* New York: Springer.

Dekker, R., & Engbersen, G. (2014). How Social Media Transform Migrant Networks and Facilitate Migration. *Global Networks, 14*(4), 401–418.

Dekker, R., Engbersen, G., & Faber, M. (2016). The Use of Online Media in Migration Networks. *Population Space and Place, 22*(6), 539–551.

Dervin, B. (1992). From the Mind's Eye of the User: The Sense-Making Qualitative–Quantitative Methodology. In J. D. Glazier & R. R. Powell (Eds.), *Qualitative Research in Information Management.* Englewood, CO: Libraries Unlimited.

Dewitz, L. (2016). *Information behavior of unaccompanied minor refugees in consideration of the role and use of smartphones* (Unpublished master's thesis). Berlin School of Library and Information Science, Humboldt-Universität zu Berlin, Germany.

Dimenescu, D. (2008). The Connected Migrant: An Epistemological Manifesto. *Social Sciences Information. Information Sur les Sciences Sociales, 47*(4), 565–579.

Fisher, K. E. (2016, June 27). Design and Field Insights with Refugees in Camps by Conflict Zones. *Open Lab, Newcastle University.* Newcastle Upon Tyne, UK.

Fisher, K. E. (2017a, February 17–18). *Beyond WiFi: How Syrian Teens Hack Futures at UNHCR Za'atari Refugee Camp.* Paper presented at the 3M Workshop of Migrants, Marginalization and Mobile, Nanyang Technological University, Singapore. Retrieved from http://www.reinform.org/3mworkshop/.

Fisher, K. E. (2017b, May 16). The Za'atari Refugee Camp Cookbook: How Space, Gender and Time affect Information Behavior. (Keynote lecture.) In *Conference on Integration in Cooperation: Developing Information Services for Refugees in Collaboration with Key Actors.* Åbo Akademi University (ÅAU), Turku, Finland.

Fisher, K. E., Borkert, M., Yafi, E., & Yefimova, K. (2017a, May 15–18). The Value of a Smart Phone and Things They'd Have Done Differently: Information Pitfalls and

Plans for the Future of Syrian Refugees in a Berlin Camp. In *10th Annual Conference of the Canadian Association for Refugee and Forced Migration Studies (CARFMS), Centre for Asia-Pacific Initiatives' Migration and Mobility Program*. University of Victoria, BC, Canada.

Fisher, K. E., Maitland, C., Yefimova, K., Xu, Y., & Yafi, E. (in press). Al Osool: Understanding People, Place and Time at Za'atari Syrian Refugee Camp.

Fisher, K. E., Omondi, I., & Yafi, E. AbdulMajeed, Z., and Yefimova, K. (2017b, May 15–18). Al Asool at UNHCR Za'atari Camp: An Asset-based Field Study of People, Place and Time. In proceedings of the *10th Annual Conference of the Canadian Association for Refugee and Forced Migration Studies (CARFMS), Centre for Asia-Pacific Initiatives' Migration and Mobility Program*. University of Victoria, BC, Canada.

Fisher, K. E., Talhouk, R., Yefimova, K., Al-Shahrabi, D., Yafi, E., Ewald, S., & Comber, S. (2017, May). Za'atari Refugee Cookbook: Relevance, Challenges and Design Considerations. In *CHI 2017*. Boulder, CO.

Fisher, K. E., & Yafi, E. (Under review, b). Syrian Youth in Za'atari Refugee Camp as ICT Wayfarers: An Exploratory Study Using Narrative Storytelling and LEGO.

Fisher, K. E., Durrance, J. C., & Hinton, M. B. (2004). Information Grounds and the Use of Need-based Services by Immigrants in Queens, New York: A Context-based, Outcome Evaluation Approach. *Journal of the American Society for Information Science and Technology, 55*(8), 754–766.

Fisher, K. E., Marcoux, E., Miller, L. S., Sanchez, A., & Cunningham, E. R. (2004). Information Behavior of Migrant Hispanic Farm Workers and their Families in the Pacific Northwest. Information Research, 10.1, paper 199. Retrieved from http://InformationR.net/ir/10-1/paper199.html.

Fisher, K. E., Yefimova, K., Dahya, N., Yafi, E., Belding, E., & Borkert, M., et al. (2016a). HCI, Forced Migration and Refugees: Collaborations Across Borders and Fields. In proceedings of *CHI 2016 Development Consortium 2016: HCI Across Borders*. San Jose, CA.

Fisher, K. E., Yefimova, K., & Bishop, A. P. (2016b). Adapting Design Thinking and Cultural Probes to the Experiences of Immigrant Youth: Uncovering the Roles of Visual Media and Music in ICT Wayfaring. In *Proceedings of the 2016 CHI Conference Extended Abstracts on Human Factors in Computing Systems* (pp. 859–871). New York: ACM. Retrieved from http://doi.org/10.1145/2851581.2851603.

Fisher, K. E., Yefimova, K., & Yafi, E. (2016c). Future's Butterflies: Co-Designing ICT Wayfaring Technology with Refugee Syrian Youth. In *IDC '16: Proceedings of the 15th International Conference on Interaction Design and Children*. Manchester, UK.

Fortunati, L., Pertierra, R., & Vincent, J. (2011). *Migration, Diaspora and Information Technology in Global Societies*. New York: Routledge.

Freedman, J. (2016, July). Moving Beyond "Vulnerability": A Gendered Analysis of the Migration "Crisis." In *16th Conference of the International Association for the Study of Forced Migration* Poznań, Poland.

Gifford, S. M., & Wilding, R. (2013). Digital Escapes? ICTs, Settlement and Belonging among Karen Youth in Melbourne, Australia. *Journal of Refugee Studies, 26*(4), 558–575.

Harris, R. M., & Dewdney, P. (1994). *Barriers to Information: How Formal Help Systems Fail Battered Women.* Westport, CT: Greenwood.

Johnston, J. (2016a). The Use of Conversation-based Programming in Public Libraries to Support Integration in Increasingly Multiethnic Societies. *Journal of Librarianship and Information Science*, February 22, 1–11.

Johnston, J. (2016b). Conversation-based Programming and Newcomer Integration: A Case Study of the Språkhörnan programat Malmö City Library. *Library & Information Science Research, 38*(1), 10–17.

Jones, K. (2016, July). "Connection men," "smugglers" and "friends": Exploring the Role of Non-State Actors who Facilitate Migration from West Africa to Italy. In *16th Conference of the International Association for the Study of Forced Migration.* Poznań, Poland.

Katz, V. S. (2014). *Kids in the Middle: How Children of Immigrants Negotiate Community Interactions for their Families.* New Brunswick, NJ: Rutgers University Press.

Koo, J. H., Cho, Y. W., & Gross, M. (2011). Coping with Severe Traumatic Stress: Understanding the Role of Information-Seeking among Political Refugees. In *Proceedings of the 2011 iConference* (pp. 699–701). New York: ACM. http://doi.org/10.1145/1940761.1940875.

Kumar, N., et al. (2017, May 6–7). HCI Across Borders Symposium. *ACM CHI Conference.* Denver, CO. http://www.hcixb.org/.

Kuschminder, K. (2016, July). Migrants' Decision Making Factors in Transit: A Comparative Analysis of Greece and Turkey. In *16th Conference of the International Association for the Study of Forced Migration.* Poznań, Poland.

Latif, N. (2012). "It Was Better during the War": Narratives of Everyday Violence in a Palestinian Refugee Camp. *Feminist Review, 101,* 24–40. doi:10.1057/fr.2011.55.

Leurs, K. (2016). Young Connected Migrants and Non-Normative European Family Life Exploring Affective Human Right Claims of Young E-Diaspora. *International Journal of E-Politics, 7*(3), 15–34.

Lingel, J. (2015). Information Practices of Urban Newcomers: An Analysis of Habits and Wandering. *JASIS&T, 66*(6), 1239–1251.

Lloyd, A. (2015). Stranger in a Strange Land: Enabling Information Resilience in the Resettlement Landscape. *Journal of Documentation, 71*(5), 1029–1042.

Lloyd, A. M., Kennan, M. A., & Thompson, A. Q. (2013). Connecting with New Information Landscapes: Information Literacy Practices of Refugees. *Journal of Documentation, 69*(1), 121–144.

Lloyd, A., & Wilkinson, J. (2016). Knowing and Learning in Everyday Spaces (KALiEds): Mapping the Information Landscape of Refugee Youth Learning in Everyday Spaces. *Journal of Information Science, 42*(3), 300–312.

Maitland, C., Fisher, K. E., Tomaszewski, B., & Xu, Y. (2016, December). Final Report to UNHCR Jordan: Information-enabled Community Engagement and Asset Mapping at Za'atari Camp. College Park, PA: Penn State University; Seattle, WA: University of Washington.

Maitland, C., Tomaszewski, B., Belding, E., Fisher, K. E., Xu, Y., & Iland, D., & Majid, A., et al. (2015). Youth Mobile Phone and Internet Use, January 2015, Za'atari Camp, Mafraq, Jordan. Pennsylvania State College of Information Sciences and Technology. Retrieved from https://cmaitland.ist.psu.edu/wp-content/uploads/sites/9/2015/01/ZaatariSurveyAnalysis2015November2.pdf.

Maitland, C., & Xu, Y. (2015). A Social Informatics Analysis of Refugee Mobile Phone Use: A Case of Za'atari Syrian Refugee Camp. In proceedings of the *43rd Research Conference on Communications, Information and Internet Policy (TPRC '15)*. Arlington, VA.

McGregor, E., & Siegel, M. (2013). *Social Media and Migration Research (No. 068). United Nations University-Maastricht Economic and Social Research Institute on Innovation and Technology*. MERIT.

Mesmar, S., Talhouk, R., Akik, C., Patrick Olivier, P., Elhajj, I. H., Elbassuoni, S., et al. (2016). The Impact of Digital Technology on Health of Populations Affected by Humanitarian Crises: Recent Innovations and Current Gaps. *Journal of Public Health Policy, 37*(Suppl. 2), 167–200.

Nardone, M., & Correa-Velez, I. (2015). Unpredictability, Invisibility and Vulnerabililty: Unaccompanied Asyylum-Seeking Minors' Journeys to Australia. *Journal of Refugee Studies, 29*(3), 1–20.

Neuenhaus, M., & Aly, M. (2017). Empathy Up. In *Proceedings of the 2017 CHI Conference Extended Abstracts on Human Factors in Computing Systems (CHI EA '17)* (pp. 86–92). New York: ACM.

Oltermann, P., & Kingsley, P. (2016, August 25). "It Took on a Life of Its Own": How One Rogue Tweet Led Syrians to Germany. *The Guardian*. Retrieved from https://www.theguardian.com/world/2016/aug/25/it-took-on-a-life-of-its-own-how-rogue-tweet-led-syrians-to-germany?CMP=share_btn_link.

Pettigrew, K. E. (1999). Agents of Information: The Role of Community Health Nurses in Linking the Elderly with Local Resources by Providing Human Services

Information. In *Exploring the Contexts of Information Behaviour* (pp. 257–276). Taylor Graham Publishing.

Quirke, L. (2012). Information Practices in Newcomer Settlement: A Study of Afghan Immigrant and Refugee Youth in Toronto. In *Proceedings of the 2012 iConference* (pp. 535–537).

Rohde, M., Aal, K., Misaki, K., Randall, D., Weibert, A., & Wulf, V. (2016). Out of Syria: Mobile Media in Use at the Time of Civil War. *International Journal of Human-Computer Interaction, 32*(7), 515–531.

Rudoren, J. (2015, September 22). As Others Flee to West, Most Syrian Refugees Remain in Region. *New York Times*. http://nyti.ms/1ONppa6.

Sampson, R., & Gifford, S. M. (2010). Place-Making, Settlement and Well-Being: The Therapeutic Landscapes of Recently Arrived Youth with Refugee Backgrounds. *Health & Place, 16*(1), 116–131. doi:10.1016/j.healthplace.2009.09.004.

Sawhney, N. (2009). Voices Beyond Walls: The Role of Digital Storytelling for Empowering Marginalized Youth in Refugee Camps. In *Proceedings of the Eighth International Conference on Interaction Design and Children* (pp. 302–305). New York: ACM (Association for Computing Machinery).

Shankar, S., O'Brien, H. L., How, E., Lu, Y., Mabi, M., & Rose, C. (2016). The Role of Information in the Settlement Experiences of Refugee Students. In *Proceedings of the 79th ASIS&T Annual Meeting: Creating Knowledge, Enhancing Lives through Information & Technology (ASIST '16) 53*(1) (pp. 1–6). Silver Springs, MD: American Society for Information Science.

Smets, K., & Leurs, K. (in press, 2018). Guest Editors. Special issue of *Social Media + Society*: "Forced Migration and Digital Connectivity in(to) Europe."

Srinivasan, R., & Pyati, A. (2007). Diasporic Information Environments: Reframing Immigrant-Focused Information Research. *Journal of the American Society for Information Science and Technology, 58*(12), 1734–1744.

Stickel, O., Hornung, D., Aal, K., Rohde, M., & Wulf, V. (2015). 3D Printing with Marginalized Children—An Exploration in a Palestinian Refugee Camp. In N. Boulus-Rødje, G. Ellingsen, T. N. Boulus-Rødje, et al. (Eds.), *ECSCW 2015: Proceedings of the 14th European Conference on Computer Supported Cooperative Work, 19–23 September 2015* (pp. 83–102). Oslo, Norway.

Takieddine, M. (2014, November 4). *Oasis of resilience: Healing and empowering Syrian children in Za'atari Refugee Camp* (Landscape architecture master's thesis). University of Washington, Seattle, WA. Retrieved from http://hdl.handle.net/1773/27114.

Talhouk, R., Bartindale, T., Montague, K., Mesmar, S., Akik, C., Ghassani, A., et al. (2017a, June). Implications of Synchronous IVR Radio on Syrian Refugee Health and

Community Dynamics. In *Proceedings of the 8th International Conference on Communities and Technologies* (C&T /17). Troyes, France.

Talhouk, R., Fisher, K. E., Wulf, V., et al. (2017b, June 26–30). Refugees & HCI workshop: The role of HCI in responding to the refugee crisis. Communities & Technologies. Troyes, France. Retrieved from https://www.researchgate.net/publication/302074231.

Talhouk, R., Mesmar, S., Thieme, A., Balaam, M., Olivier, P., Akik, C., et al. (2016). Syrian Refugees and Digital Health in Lebanon: Opportunities for Improving Antenatal Health. In *ACM Conference on Human Factors in Computing Systems* (pp. 331–342).

Toyama, K. (2015). *Geek Heresy: Rescuing Social Change from the Cult of Technology*. New York: Public Affairs.

Turner, S. (2015). What Is a Refugee Camp? Explorations of the Limits and Effects of the Camp. *Journal of Refugee Studies, 29*(2), 139–148.

UNICEF/REACH Initiative. (2014). Access to education for Syrian refugee children in Za'atari Camp, Jordan. Retrieved from http://www.unicef.org/jordan/Joint_Education_Needs_Assessment_2014_E-copy2.pdf.

United Nations Children's Fund Regional Office for the Middle East & North Africa. (2015). Education under fire: How conflict in the Middle East is depriving children of their schooling. New York: United Nations Children's Fund. Retrieved from http://www.unicef.org/mena/media_10557.html.

United Nations Department of Economic and Social Affairs. (2013). Definition of Youth. Retrieved from http://www.un.org/esa/socdev/documents/youth/fact-sheets/youth-definition.pdf.

United Nations High Commissioner for Refugees. (2015a). UNHCR Refugee Resettlement Trends 2015. Retrieved from http://www.unhcr.org/559e43ac9.html.

United Nations High Commissioner for Refugees. (2016d, February 26). UNHCR, UNICEF launch Blue Dot hubs to boost protection for children and families on the move across Europe. Retrieved from http://www.unhcr.org/news/press/2016/2/56d011e79/unhcr-unicef-launch-blue-dot-hubs-boost-protection-children-families-move.html.

United Nations High Commissioner for Refugees. (2015b, July 9). Total number of Syrian refugees exceeds four million for first time. Retrieved from http://www.unhcr.org/559d67d46.html.

United Nations High Commissioner for Refugees. (2016a). Global Trends: Forced Displacement in 2015. Geneva, Switzerland. Retrieved from https://s3.amazonaws.com/unhcrsharedmedia/2016/2016-06-20-global-trends/2016-06-14-Global-Trends-2015.pdf.

United Nations High Commissioner for Refugees. (2016b). Missing Out: Refugee Education in Crisis. Geneva. Retrieved from http://www.unhcr.org/57d9d01d0.

United Nations High Commissioner for Refugees. (2016c). Trends at a glance: 2015 in review. Retrieved from https://s3.amazonaws.com/unhcrsharedmedia/2016/2016-06-20-global-trends/2016-06-14-Global-Trends-2015.pdf.

United Nations High Commissioner for Refugees. (2016d). *3RP 2016 Mid-Year Report.* Retrieved from http://www.3rpsyriacrisis.org/wp-content/uploads/2016/07/3RP-Mid-year-Report.pdf.

Vernon, A., Deriche, K., & Eisenhauer, S. (2016). Connecting refugees: How Internet and mobile connectivity can improve refugee well-being and transform humanitarian action. Geneva: United Nations High Commissioner for Refugees. Retrieved from http://www.unhcr.org/en-us/publications/operations/5770d43c4/connecting-refugees.html.

Wendle, J. (2016, January). For asylum seekers, a cellphone is a bridge to the future and the past. *Al Jazeera.* Retrieved from http://america.aljazeera.com/articles/2016/1/18/for-asylum-seekers-a-cell-phone-is-a-bridge-to-the-future-and-the-past.html.

Yafi, E., Nasser, R., & Tawileh, A. (2015). ICT's Impact on Youth and Local Communities in Syria. In *ICTD '15: Proceedings of the Seventh International Conference on Information and Communication Technologies and Development* (pp. 68:1–68:4). New York: ACM. doi:10.1145/2737856.2737907.

Yafi, E., Yefimova, K., & Fisher, K. E. (in press). Young Hackers. *Hacking Technology at Zaatari Syrian Refugee Camp.*

Yerousis, G., Aal, K., von Rekowski, T., Randall, D. W., Rohde, M., & Wulf, V. (2015). Computer-enabled Project Spaces: Connecting with Palestinian Refugees Across Camp Boundaries. In *Proceedings of the 33rd Annual ACM Conference on Human Factors in Computing Systems* (pp. 3749–3758). New York: ACM. Retrieved from http://doi.org/10.1145/2702123.2702283.

II Technical Perspectives

6 Cellular and Internet Connectivity for Displaced Populations

Paul Schmitt, Daniel Iland, Elizabeth Belding, and Mariya Zheleva

Network connectivity, Internet or cellular, is a critical need for both displaced people and the aid agencies that serve them. Not only do displaced people need to remain in contact with family they have left behind, they must reconnect with missing kin who may have been lost during relocation. While in normal circumstances, network connectivity is provisioned through a combination of fixed, wireless, and cellular networks, in displacement scenarios, connectivity can be most plausibly resolved through robust and reliable cellular service.

Unfortunately, cellular communication services and infrastructure can be nonexistent in areas where displaced people settle, and where they do exist, they often provide poor service. To better understand the dimensions of service quality, in this chapter we examine real-world communications infrastructure performance using traces collected in an operational refugee camp. We find that the commercial cellular infrastructure providing service in the camp often suffers from radio resource congestion, which manifests in the inability for users to obtain voice/SMS or data service in a timely manner. Next, we discuss current networking system solutions that may be applicable to displacement scenarios. Lastly, we outline a research agenda, including both technical and policy dimensions, for improving connectivity for those affected by displacement. We argue that policies for occupying wireless frequency spectrum must allow for increased sharing of this finite resource, and corresponding technologies that enable such interaction must be pursued in this context. We also assert that bottom-up, user-extensible communication infrastructure is critical to bridge the connectivity gap currently facing displaced populations.

The Broader Context of Wireless Connectivity for Displacement

UNHCR, the United Nations Refugee Agency, specifically recognizes that cellular connectivity is crucial for both the empowerment of displaced communities and improved humanitarian assistance (Vernon, Deriche, & Eisenhauer, 2016). Unfortunately, displaced people often live in areas that offer either unreliable connectivity or no connectivity options at all. Twenty percent of refugees living in rural areas have no cellular access, and only 17% of rural refugees live in areas covered by infrastructure offering mobile broadband speeds (i.e., 3G or better), compared with global rural population coverage of 29% (ITU, 2015). Refugees also recognize the critical importance of connectivity, spending up to one third of their disposable income on access. Lastly, online content and services are often irrelevant for displaced populations (e.g., non-native language), and digital literacy training is often necessary as refugees that have had limited exposure to the Internet can have difficulty learning to use it. Accordingly, the UNHCR is actively pursuing a multi-pronged agenda to address availability, affordability, and usability challenges with respect to the populations they serve. Principally, connectivity, via cellular and fixed infrastructure, must first be deployed before affordability and usability can be addressed.

In addition to humanitarian agencies responsible for protection and support of displaced people, the private sector is working to increase connectivity options by exploring alternative infrastructures. Recently, statements by Facebook CEO Mark Zuckerberg indicating his intention to provide Internet connectivity to refugee camps (Sengupta, 2015) have yet to generate concrete and targeted outcomes. Instead, it appears they are part of broader efforts to expand connectivity through the organization Internet.org. With goals to expand connectivity, Facebook has explored novel ways for providing connectivity in rural, unconnected areas using drone[1] technologies. However, this solution is designed to provide middle-mile connectivity to an area with the expectation that end-users will connect through WiFi or some other local access technology. For last-mile connectivity, this effort has thus far aimed to provide data, not cellular service. Given the continued ubiquity of feature phones, the availability of voice and SMS services, in addition to data connectivity, remain critical for displaced people. Further, cellular technologies can serve a much larger footprint compared with WiFi (35km distance limit versus hundreds of meters for WiFi, typically). For these reasons cellular service offers the most scalable, effective, and universal means of connectivity. However, WiFi can certainly be used to complement cellular, and can be a particularly useful solution for key access points such as education centers, when it is available. Through Project Loon, Google has explored the use of balloons[2] to extend cellular connectivity to rural, unconnected areas. This project relies on partnerships between cellular operators and Google to relay nearby LTE signals to users. While smartphones are increasingly popular, basic feature phones (i.e., phones typically lacking LTE radios) will continue to represent a sizable portion of the market[3] for the foreseeable future. Such phones are incompatible with Google's solution.

The need for connectivity is recognized by aid agencies, which have created programs to help the displaced gain access to cellular technology in order to maintain contact with the outside world.[4] In camps, which are often far from population centers, the success of these programs depends upon the availability of reliable cellular access. Unfortunately, cellular services in refugee camps are often unavailable, spotty, or of poor quality at best. The conflicts that caused the displacement often result in the destruction of telecommunication infrastructure. Further, both planned and unplanned camps typically are established in undeveloped areas, leading to a concentration of people relying on a rural cellular infrastructure that was never provisioned to support the sheer number of requests posed by a nearby camp, if that infrastructure exists at all. Such changed utilization is likely to render any remaining communication equipment unusable due to traffic overloads. Further complicating matters, displaced people often have little money, either because circumstances required they leave it behind, or because the infrastructure to access their money is not available. The lack of and the poor quality of technological infrastructure, coupled with the limited access to finances, present seemingly insurmountable obstacles to communication. Yet it is in exactly these circumstances that communication is most critical. Further, due to the limited financial resources of displaced persons, and the desire for refugee camps to remain "temporary," existing cellular infrastructure is unlikely to be upgraded to handle the increase in load created by the new camp residents.

In addition to refugees' own communication needs, relief organizations often desire to communicate with camp residents through cellular networks, i.e., through camp-wide SMS broadcasts or personal phone calls.

The lack of cellular infrastructure and the frequent overloading of existing infrastructure can be sources of frustration for camp administrators, resulting in the inability to ensure reliable communication with camp residents and between administrators themselves. This chapter explores the current real-world performance of wireless infrastructure serving a displaced population, and applicable networking systems research that can be leveraged to improve connectivity in such scenarios.

Real-World Camp ICT Performance

In order to design connectivity solutions, we must first understand the realities faced by users of camp-serving infrastructure. The establishment of refugee camps or unplanned settlements, often in previously rural areas with scant infrastructure, frequently leads to the collapse of extant cellular networks due to the unforeseen increase in user demand. This results in information and communication technologies (ICTs) serving settlements and camps simply being unable to provide adequate connectivity for residents. An example was explored in Schmitt, et al. (2016a), where community-level digital divides were revealed in the Za'atari Syrian refugee camp in northern Jordan. We rely on this analysis to provide insight into the operational capacity of ICTs in real-world scenarios. As compared to most, Za'atari is relatively well-connected and is served by three commercial cellular carriers. Still, the infrastructure fails to meet the demands for bandwidth and connectivity, for refugees and the service providers operating in the camp alike.

In their analysis of the camp's cellular infrastructure, the authors leverage GSM control messages, using them as proxy measures for base station control channel congestion. Figure 6.1 displays the message sequences that take place when a phone or mobile station (MS) requests a private communication channel from the base station (BTS), which is necessary for both voice/SMS or data. Messages broadcast from the base station to the mobile phone (solid arrows in figure 6.1) can be captured, since they are publicly available, and analyzed to infer local system congestion. When a phone needs to use the network for a voice/SMS or data session, it issues a channel request to the base station. If a channel is available, the base station responds with an *immediate assignment success* message that provides information about the available channel. A base station operating at

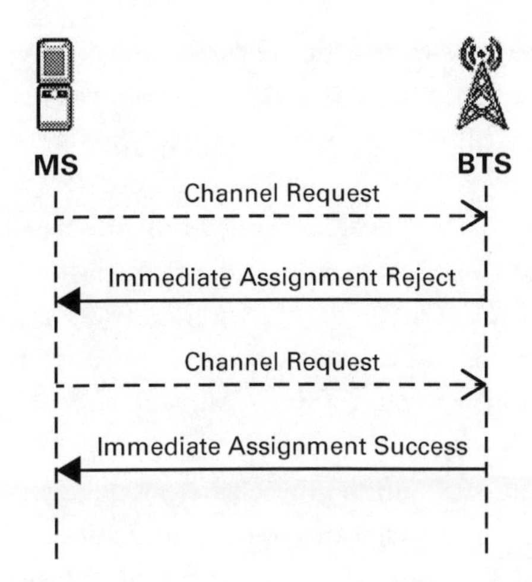

Figure 6.1
GSM network immediate assignment procedure. Broadcast messages from the base station (BTS) to the phone (MS) are captured over the GSM air interface.

full capacity such that it is unable to allocate a channel will issue an *immediate assignment reject* message, indicating no channel is available. Because these control messages are broadcast to all phones connected to the base station, they can be leveraged to approximate what is known as "standalone dedicated control channel" (SDCCH) blocking (Haider, Zafrullah, & Islam, 2009; Kyriazakos, Karetsos, Gkroustiotis, Kechagias, & Fournogerakis, 2001), which is a standard measure of base station congestion. The immediate assignment success rate is calculated by dividing the number of observed successful immediate assignment messages by the total number of immediate assignment (success and reject) messages.

Research from Za'atari shows the three cellular carriers serving the camp have significantly different success and rejection rates. Zain, the most popular carrier with camp residents, experienced the most sustained congestion, occurring throughout the day and frequently reaching rejection percentages above 60%. The Orange network was congested in short, severe bursts, likely coinciding with camp staff workday schedules (i.e., after 10:00 and before 14:00). Umniah, on the other hand, exhibited almost no evidence of congestion throughout the day.

A second indicator of congestion available through analysis of immediate assignment rejection messages is the *backoff wait value,* indicating how long a phone must wait until it repeats its request for a resource (emergency calls are not subject to the wait value restriction). Cellular networks use the backoff wait value to ease congestion caused by phones repeatedly requesting unavailable resources in quick succession. The value ranges from 0–255 seconds, with the value determined by the level of overload on the base station. This value can be leveraged to indicate the *severity* of congestion. Measurements from Za'atari indicate Zain had many 128-second waits, an indication of significant congestion. In these instances, a user must wait roughly two minutes before reattempting to place a voice call, send an SMS, or use data services.

A third indication of service quality, in addition to accept/reject messages and backoff wait times, is the type and speed of connection achieved, namely whether 3G HSPA or 2G Edge connections are available. With measurement phones set to prefer 3G connectivity, the measurements indicated Zain's 3G connectivity was mostly focused around the northern road, where UN and Jordanian security offices are located, along with a few small areas around the perimeter of the camp. Both Zain and Orange had areas with no data connectivity in the north central area of the camp; Zain also lacked data coverage in the southwest area of the camp. Orange had the smallest area of 3G coverage, with the majority of the camp covered by 2G EDGE connectivity. Umniah, the least popular carrier according to residents surveyed, had the highest number of 3G coverage points.

Different Timescales, Different Connectivity

The camp measurements clearly indicate the commercial cellular infrastructure is overburdened and unable to handle the client load. These results provide empirical support to residents' reports of widespread outages and unpredictable connectivity throughout the day (Pizzi, 2013). These findings also underscore the challenge of rapidly deploying connectivity in response to sudden usage changes. Generally, traditional network infrastructure and its management are not designed for rapid expansion.

To summarize, and drawing upon insights from the measurements above, it appears that the current capabilities of infrastructure serving displaced populations varies across time with respect to the displacement event, as illustrated in figure 6.2. Immediately after displacement, carriers are often limited to providing basic SMS and voice communications, only providing

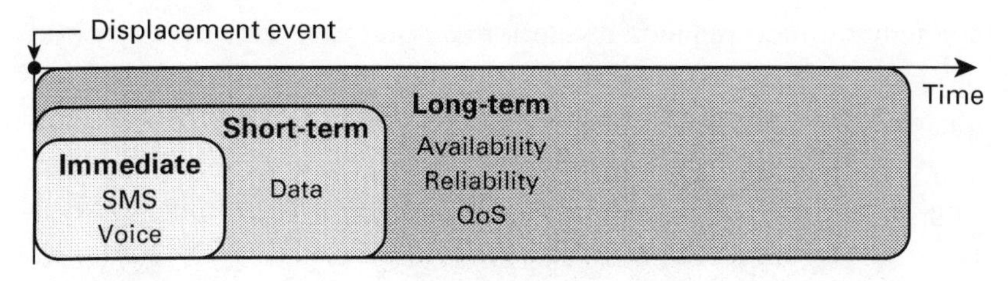

Figure 6.2
Current displacement networking capabilities vary with time.

more advanced data services in the short- and long-term. While SMS and voice can provide a vital lifeline for displaced individuals, immediately, during, and after displacement, affected users need more advanced services, such as data access. Such services enable refugees to contribute crowd-sourced information through platforms, such as Ushahidi,[5] or to post status updates on social networks. Further, some displaced populations, such as refugees, may have to spend a substantially long period of time in their new location. It is often only in these long-term cases that carriers are able to shift focus to increasing the reliability and quality of service of the provided connectivity. The limitations and technical bases of this current "service trajectory" are discussed below.

Immediate Immediately after a displacement event, the communication infrastructure is likely to suffer poor performance or be absent all together. At the same time, the communication needs of displaced populations are most critical in the hours and days immediately after displacement, which often results in a surge in demand on existing communication infrastructure. Since voice and SMS put less burden on the communication channel, the immediate surge demand can be accommodated by less, simpler infrastructure. Furthermore, SMS is delay-tolerant, which allows for use of delay-tolerant backhaul (such as drones) while traditional real-time backhaul is restored.

Short-term In the days and weeks following a displacement, the performance of existing communication infrastructure may begin to improve as this infrastructure is repaired and backhaul access recovered. At the same time, user behavior begins to change in terms of demand. The initial surge of communication requests becomes more uniform over time, which in turn puts less strain on the network and allows it to handle more diverse user requests.

Long-term Some displaced populations, such as refugees, may need to stay in their new location for long periods of time. In such cases the communication needs of these populations begin to resemble the day-to-day communication activity of non-displaced users. The offered load becomes more regular and predictable, and the demand is a mix of voice, SMS, and data services. The requirements posed to the network infrastructure are high availability and reliability, as well as certain Quality of Service (QoS) guarantees. Carriers are able to deploy infrastructure over longer time-scales, tailored to serve the community.

An ideal system needs to be able to intelligently sense the post-displacement phase and accurately allocate resources to be able to best handle the user demand. For example, such systems should be able to characterize the user demand and adaptively allocate resources across users. It should also be able to sense and characterize the supply, i.e., the availability and quality of backhaul and Internet access, and adapt its service portfolio accordingly.

System Solutions

Depending on the displacement context, the need for alternative communication infrastructure may be very rapid and short term, e.g., in post-disaster scenarios, or medium- to long-term, as in the case of refugee camps. In either case, there is a need for communication solutions that are highly robust, self-contained and rapidly deployable. Such solutions should provide at the very minimum basic services, including text messages and voice calls, and ideally, data services as well. Recent developments in this space have all adopted small local cells as a basis for design of infrastructure for displaced persons. In this section, we first detail the underlying technology of small local cells and describe several practical deployments in this space. We then survey recent advances that harness small local cells to provide connectivity to displaced populations in a spectrum of scenarios ranging from post-disaster to refugee camps.

Local cellular networks Small local cellular networks have been proposed in recent years to provide connectivity in infrastructure-challenged areas

(Heimerl & Brewer, 2010; Heimerl, Hasan, Ali, Brewer, & Parikh, 2013; Rhizomatica, 2015; Zheleva, Paul, Johnson, & Belding, 2013). Small local cellular networks make use of open-source software and open hardware in order to create technology that: (1) is fully compatible with users' existing phones and SIM cards; (2) does not require commercial-grade backbone; and (3) has low infrastructure needs. As such, these networks can efficiently utilize renewable/alternative power sources, harness generic IP networks for backbone connectivity, and provide mobile cellular services at a fraction of the capital and operational expenditures incurred by commercial providers. Thus, small local cells are well-positioned to meet the key technology requirements for displaced persons outlined in the beginning of this chapter.

A typical architecture of small local cells is depicted in figure 6.3 (Zheleva, Paul, Johnson, & Belding, 2013). The base stations run OpenBTS, which implements the GSM stack and communicates with the associated cell phones using the standard Um radio interface for commodity 2G and 2.5G cell phones. OpenBTS is also responsible for translating GSM messages to SIP, which allows the use of low-cost generic IP backbone infrastructure, as opposed to an expensive commercial-grade GSM backbone. SIP translation enables the use of free VoIP server software to serve as a Mobile Switching Center for routing calls.

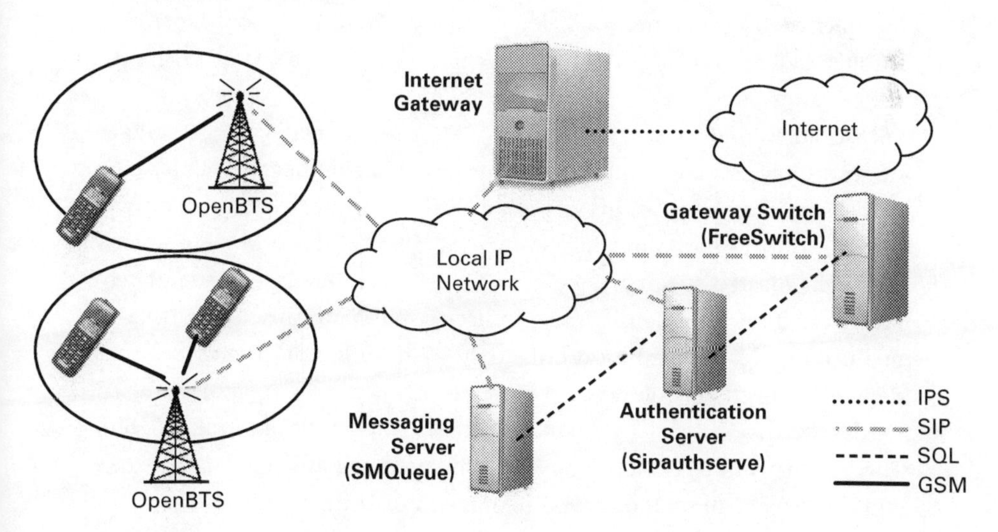

Figure 6.3
Local cellular network architecture.

FreeSwitch,[6] an open-source private branch exchange (PBX), is used to route calls within and outside the network. FreeSwitch connects to OpenBTS via SIP and RTP, and routes calls both in intra- and inter-BTS local scenarios. It has the capability to route calls outside of the network to commercial cellular, fixed line, and VoIP networks using SIP and SS7. By the means of custom Python scripts, FreeSwitch allows extension of the basic routing functionality to facilitate cell phone based applications.

The network utilizes Sipauthserve and SMQueue to handle user authentication and text messaging, respectively. SMQueue is the SIP-based equivalent of an SMSC (Short Message Service Center) in a commercial-grade system. As such, it interfaces with OpenBTS and makes use of commodity IP networks to transmit SMS (Short Message System). At the same time it can interface with commercial SMSCs using SS7 and SMPP. SMQueue implements a store-and-forward SMS queue functionality that allows messages to be delivered in a delay-tolerant fashion. This last functionality is of great importance for areas with intermittent cell phone access and electric power availability, since users are often either out of range or have their cell phones powered off. To handle user authentication and mobility, the network leverages Sipauthserve—a database server with an interface to process SIP REGISTER messages to track mobility. Both SMQueue and Sipauthserve are queried by other network elements (e.g., FreeSwitch and OpenBTS) through SQL.

Local cellular for displaced populations Prior deployments of local cellular networks have focused on areas with no existing commercial cellular coverage. Displaced populations present a different use–case, as increasingly, with the further build-out of commercial cellular networks, settlements are often located in areas where coverage exists but is not adequately provisioned for a sudden increase in user population caused by displacement. Building on small local cells, there is an influx of recent work that designs connectivity solutions to meet displaced persons' communication needs (Ghaznavi, et al., 2014; Iland & Belding, 2014; Schmitt, et al., 2016b). These systems have explored rapid deployment of local cellular networks and coexistence of such networks along with existing, overburdened commercial networks.

To address the communication needs of residents and aid agencies in refugee camps, there is a need for infrastructure alternatives that: (1) allow for rapid deployment; (2) do not require specialized backbone infrastructure (3) can operate off the power grid with low power consumption; and (4) can function in coexistence with, and complement the services provided by, commercial communication networks. We envision that the next step will be "hybrid" cellular networks, where local networks augment and coexist with commercial cellular networks in a location where commercial coverage is unable to meet demand. We envision the following applications for such systems:

- **Local services for end users.** Due to the high spatial locality of interest of cellular communications, local networks can provide low-cost local cellular services to users within the camp area. Furthermore, if Internet access is available, a local network can provide outbound services. Where the commercial cellular network is still available, users can choose whether to use the local or the commercial cell for outbound communication.
- **Local services for organizations.** Local networks can also be used by different organizations, such as camp management or schools, for rapid outreach to large user populations. Applications can be developed that make use of local cellular networks to improve the information channels between service providers and local residents.
- **Added services.** Where the commercial network service profile has been reduced to voice and SMS services only, local cellular systems can provide for data access. Even where data services are available through the commercial provider, in the face of a mobile data surge, the local cell can be used for mobile data offloading.

A variety of research challenges arise related to the operation of such a hybrid cellular network. First, although both the commercial and local networks will provide service to users, these two networks need to be perceived as a single entity by the end user. Second, challenges exist related to the seamless utilization of two individual networks with users' existing phones and SIM cards. For this seamless operation, the local cell needs to be cognitively aware of the operational state and capabilities of the commercial cell. This requires hardware extensions for continuous spectrum sensing, as well as algorithms to mine the spectrum traces and infer the commercial cell's activity. Finally, the ability to provide a union of services via two individual networks poses unique challenges in application-aware handover among multiple cellular networks.

Research Agenda

So far, we have summarized several promising system designs that alleviate the communication problems in the displaced persons context. In this section, we identify several limitations of the current work and outline an agenda for future developments in this space.

In order to unleash the full potential of these systems to improve the communication opportunities of displaced populations, there is a need for both technological improvements as well as fundamental advances in policy. In terms of technology, while the current solutions discussed in the System Solutions section meet the most immediate needs of text and voice communications, they lack the ability to provide data services, and their ability to provide global connectivity is dependent on backhaul availability. The policy domain poses even more challenging problems. Current spectrum regulations constrain the operation of small local cells by requiring expensive spectrum licenses. Furthermore, in some displaced contexts, such as the refugee case, host countries are reluctant to provide communication infrastructure to refugees.[7]

These challenges shape a compelling research agenda that calls for design of technology that brings a full set of services to displaced persons, and allows autonomous infrastructure deployment. To this end, integration of data services in the existing architectures is essential. This brings new challenges in network characterization and in seamless transition of data sessions across multiple non-coordinated carriers.

To advance policy, there is a need to rethink spectrum regulations in order to depart from the current monopoly-based spectrum ownership model and allow opportunistic spectrum reuse. The USA is paving the way in this space through recent regulations in the 3.5 GHz domain that allow for a three-tiered model of spectrum reuse. At tier–1 of this model are the federal incumbents and fixed satellite service who have been traditionally licensed to these bands and will continue to have complete interference protection from the other two tiers. At tier–2 operate the Priority Access (PA) licensees, who gain priority access at a geographical location for a three-year term. This shorter term, relative to our current spectrum management practices, allows for a quicker reallocation of frequencies that are not actively used by PAs. The tier–3 operators in this model are called General Authorized Access (GAA) operators and are allowed at any geographical location to use up to 80 MHz continuous spectrum that has not been assigned to higher-tier providers. This model will allow opportunistic access of non-primary users to spectrum, where the incumbents are not active, and will alleviate the burden of licensed operation of experimental devices in rural areas. Such access brings challenges in spectrum sensing and characterization as well as network operation under uncertain spectrum availability.

Lastly, future research needs to address the need for autonomous deployment and operation of communications infrastructure. This problem is relevant both in short-term (post-disaster) and mid-to-long-term deployments (refugee camps). Current solutions require dedicated equipment (e.g., OpenBTS base stations) to establish a network. In cases where such equipment is not readily available, there is a need for self-deploying and self-organizing networks that use readily available equipment, such as end users' smartphones. With recent advances in smartphone interfaces, there is potential for development of novel infrastructures that harness smartphones to create blanket coverage in a given area. Such networks can also handle local message distribution. Where commercial service is available, these mesh networks can make use of their aggregate capacity to relay messages outside of the network. Where commercial coverage is not available, backhaul can be established through more delay-tolerant means, for example, based on unmanned aerial vehicles. The following sections identify both specific spectrum policy and technical aspects, bottom-up and infrastructure-less, of a research agenda to foster reliable network connections for the displaced.

Spectrum Regulation

An often-overlooked challenge facing wireless communications systems that operate within licensed bands is due to spectrum regulations. For example, cellular systems operate over licensed frequencies; carriers obtain exclusive rights to portions of the radio spectrum in specific locations. These agreements are necessary to protect the licensees (providers) from wireless interference that can degrade services and cause communication failures. In densely populated areas, spectrum licensing is essential for providing predictable wireless connectivity. For instance, the 900 MHz frequency range used by GSM includes a total of 125 non-overlapping wireless channels. As GSM base stations have a maximum range of 35 km, providers in densely populated areas have a clear need for strict spectrum licensing enforcement;

without it they would quickly exhaust the available frequencies and interference due to frequency reuse would degrade performance.

However, the scale of spectrum licensing rights, which is often at the national level, has an unintended consequence for coverage in rural areas. In the vast majority of countries, there is currently no legal means to deploy wireless infrastructure that operates in licensed frequencies without owning the corresponding spectrum licensing rights, despite the fact that in many rural locations the licensed spectrum is relatively unused. Additionally, spectrum regulation has been slow to evolve compared with the rapid advancement of wireless technologies.

Whereas the traditional policies and regulations are inflexible, new radio technologies, such as software-defined radios (SDRs), open the possibility of agile frequency use. Using SDRs, we can sense a given frequency and determine whether it is occupied by a carrier. We can also flexibly operate wireless systems across broad ranges of frequency spectrum. Spectrum regulators have begun to react to the technological advancements by proposing new spectrum occupancy and licensing models, such as licensed shared access (LSA), authorized shared access (ASA), and the three-tiered model previously discussed. These new models allow for spectrum to be used by more than one entity within a regulatory environment. Trials have been conducted using the new sharing models in Europe and show promise (Palola, et al., 2014; Matinmikko, et al., 2015). Essentially, primary licensees continue to have exclusive rights to spectrum they have purchased; however, secondary providers can operate in licensed spectrum in locations where the primary licensee does not use it. LSA and ASA represent major steps forward for small-scale wireless system operation. Unfortunately, the bands of spectrum where sharing has so far been allowed are relatively small, and, in the 3.5 GHz case, largely unusable outside of the U.S., and, additionally, require new phone radio hardware that is not yet widely available. Moreover, the pace at which new spectrum licensing models are actually put into place is dictated by the speed of governments.

To identify new spectrum for dynamic spectrum access (DSA), the U.S. government, industry, and spectrum regulators worldwide have endeavored to create a large-scale spectrum inventory in order to determine spectrum usage at different locations over long periods of time. For example, in the U.S., the goal of the Spectrum Inventory Bill (Congress, 2009) is to create a nationwide footprint of spectrum usage over time. Based on these measurements, spectrum regulators can open new portions of the spectrum for DSA (Chowdhery, Chandra, Garnett, & Mitchell, 2012). Furthermore, new DSA technologies can be designed taking into account the characteristics of these bands. Creating such national spectrum inventory is aimed at answering various questions (Nandagopal, 2016), including: (1) How much spectrum is occupied and how much is idle? (2) How many transmitters occupy a given frequency band? (3) Are they authorized to operate in this band? While the first question can be approached by simple estimation of power level in a given band, the other two questions require more elaborate analysis of spectrum occupancy. Such analysis needs to answer questions such as: Is there more than one transmitter in a given band? And what are their received powers, operating frequencies, bandwidth, and temporal characteristics? Learning these characteristics from raw spectrum measurements is critical for improved policing and technological advancements in the DSA domain.

Despite the need for deep understanding of spectrum occupancy, there does not exist a platform to create such a nationwide spectrum usage footprint. This is primarily due to lack of scalable infrastructure for collection and processing of RF spectrum measurements. Traditionally, spectrum occupancy is analyzed via spectrum analyzers that capture large amounts of data. The latter poses challenges in scalable data storage. Furthermore, the current approaches to mining and summarizing spectrum measurements are very limited, making it hard to evaluate the collected spectrum data. Research into accurate spectrum sensing and characterization is needed in order to actualize spectrum sharing systems that successfully protect primary carriers from interference caused by secondary carriers. We also must account for heterogeneity in spectrum sensing hardware and software. Lastly, we must explore non-interfering, uncoordinated systems that occupy the same spectrum.

Another promising research direction is through the exploration of cellular connectivity within *unlicensed* frequency bands. Recently, the telecommunications industry has begun developing hardware that operates in unlicensed ISM bands (e.g., WiFi frequencies). Carriers aim to take advantage of a new mechanism available in the LTE specification known as carrier aggregation, where multiple LTE carrier channels are logically bound to appear as a single link on the mobile device. User devices utilizing carrier aggregation can connect to multiple LTE base stations, in both licensed cellular bands and unlicensed bands, and bind all of the available channels logically as one large channel, thus increasing available bandwidth over heterogeneous networks. Unfortunately, the primary carriers that are driving the research in this space have expressed the intent to use only carrier-approved base stations that operate in unlicensed spectrum, so that they have full control over the quality of service and deployment of unlicensed cellular.

This top-down, carrier-controlled model is shortsighted. Mobile traffic is shifting toward the use of the IP multimedia subsystem (IMS) architecture, meaning call services such as voice and SMS are placed over packet-switched networks rather than traditional circuit-switched networks. This transition, as well as the rise in processing power of mobile devices, provides the opportunity to rethink mobile connectivity. Today, more and more smartphones are capable of operating their own voice-over-IP (VoIP) software directly on the device. This means that the voice or SMS traffic exiting the phone can simply be data packets, and any Internet connection will suffice for connectivity. In fact, voice-over-LTE (VoLTE) is a new standard that is beginning to be rolled out by some carriers and is often touted as an upgrade to "high-definition" voice because the codecs used can take advantage of large data bitrates for high fidelity audio.

There are many research challenges related to the use of unlicensed frequencies for cellular. Most importantly, the wireless channel is shared with uncooperative devices, resulting in unpredictable interference that makes quality of service and service guarantees difficult to provide. An uncoordinated channel may require loosening of strict timing requirements for cellular devices. Likewise, utilization of multiple data paths across heterogeneous networks presents issues related to unequal carrier properties (e.g., loss, latency, throughput). As such, intelligent selection of paths by the client based on application requirements may prove necessary in order to reach acceptable performance. Lastly, the use of cellular within the shared wireless channel should not disproportionally impact existing channel users (e.g., WiFi). Cellular systems must allow for "fair" access to shared bands.

Bottom-Up Infrastructure

As presented in the System Solutions section, locally deployable cellular infrastructures present an interesting research direction for providing connectivity to displaced people. This bottom-up approach to infrastructure deployment represents a fundamental shift in the way connectivity is provided.

To know the challenges facing new infrastructure deployments, we must understand the current industry model for providing connectivity. Traditionally, cellular and Internet connectivity deployments are based around centralized, top-down infrastructure deployed by major telecommunications companies. The inertia of this reality makes fundamentally altering the delivery of connectivity challenging, particularly for those who are not a part of the controlling organizations. In many countries, telecommunications markets have evolved over time to resemble monopolies or duopolies. Such oligopolies render it nearly impossible to deploy small-scale wireless infrastructure without the cooperation of a major service provider. For instance, for any local-scale cellular network to be viable in the long-term, it must be connected to the global PSTN.

Unfortunately, obtaining interconnection agreements from larger cellular carriers has proven difficult for local cellular networks in the past (Heimerl, Hasan, Ali, Brewer, & Parikh, 2013). Enabling bottom-up infrastructure deployment opens interesting potential avenues for future research. For instance, could cellular networks adopt a tiered model similar to Internet ISPs, in which major carriers (tier–1 providers) offer universal peering and interconnection while lower-tier providers purchase transit from other providers and offer small-scale cellular service (i.e., local cellular network)?

Up to this point, the barriers to such a model have been due to the hierarchical nature of cellular networks as cellular cores typically operate few entities that manage registration, accounting, authentication, and ingress/egress onto external networks. The introduction of LTE changes the structure of the cellular core, pushing many functions to the network edge. In theory, this flattening of the network should enable innovation at the local level as base stations can offer service in a more decentralized manner. Future work on providing a universal interconnect for voice/SMS/data services could enable a new paradigm in cellular network infrastructure.

Recent work has deployed local cellular networks (Heimerl & Brewer, 2010; Heimerl, Hasan, Ali, Brewer, & Parikh, 2013; Zheleva, Paul, Johnson, & Belding, 2013) in remote areas without existing service. Successful deployments by nonprofit and community organizations show that it is technologically and economically sustainable for community groups to deploy and operate their own "pop-up" cellular networks, which can quickly expand

and shrink in concert with demand. We identify multiple research directions for local cellular networks moving forward: (1) non-cooperative coexistence between local and commercial cellular networks in areas where coverage overlaps; (2) data onloading, that is, shifting data traffic from the commercial to the local cellular network; and (3) communications and content localization.

In areas where commercial coverage exists but is inadequately provisioned for the user population, local cellular networks may offer supplemental capacity to users, easing the burden on the commercial network. A challenge behind such operation is that local infrastructure is not recognized by the commercial network and must not interfere or degrade the existing commercial service. Data onloading between local and commercial carriers poses challenges in simultaneous characterization of multiple carriers and efficient transition across carriers that maximizes throughput while minimizing overhead. Key tradeoffs to consider are improved user experience versus additional disconnect time and battery expenditure introduced by network transition. Furthermore, beyond challenges at the physical layer transition, there is a set of problems related to seamless movement of user sessions across non-coordinated networks that require reconceptualizing conventional network protocols and infrastructures. Lastly, prior studies have found strong locality of interest in communications patterns as well as Internet content consumption (Johnson, Pejovic, Belding, & van Stam, 2012; Schmitt, Raghavendra, & Belding, 2015; Vigil, Rantanen, & Belding, 2015). Local cellular networks are well suited to take advantage of such locality as they essentially move network core functionality to the client-facing edge; client traffic between local users does not require upstream connectivity and locally produced data content can be readily placed at the network edge.

Infrastructure-Less Communication

While bottom-up infrastructures would provide a great benefit in challenging network environments, future research should also focus on fundamental connectivity alternatives. Currently, cellular and Internet infrastructure is most often delivered by specialized, expensive equipment. The high investment costs of infrastructure lead to providers being hesitant to deploy infrastructure in locations they view as temporary, a common characteristic of settlements of displaced people. We must reconsider the equipment and assumptions we currently use to deploy network infrastructure. Ideally, connectivity could be achieved using equipment that is readily available and portable. For instance, smartphones have become nearly ubiquitous and include increasingly powerful capabilities. A promising forthcoming feature that is included in the next generation of LTE standards is known as LTE-Direct. Using LTE-Direct, user devices will be able to send SMSs and place voice calls directly between user devices without the use of a cellular base station, provided the users are near one another. We believe that the functionality offered by LTE-Direct can provide a foundation for meshlike cellular networking, where nearby LTE-Direct phones not only can offer direct communications between peers, but also act as gateways to global cellular connectivity, provided they are in range of traditional cellular coverage. Meshlike cellular introduces many interesting research challenges: (1) predictable quality of service; (2) routing protocols and metrics pertaining to the desirability of paths; and (3) management of infrastructure mobility rather than simply user mobility.

Quality of service (QoS) is critical for voice traffic as it is latency-dependent. Traditional infrastructure manages QoS by strictly controlling user access to infrastructure resources. In meshlike networking models, access to the network is decentralized, opening the possibility for oversubscription of resources. In order for meshlike cellular to provide adequate service guarantees, coordination of client access is required. Clients using meshlike networks often have multiple paths available, of varying lengths, to reach gateway devices. Routing protocols operating on such a network will require metrics that account for the unique requirements of cellular traffic. Lastly, peer devices acting as infrastructure cannot be expected to be permanent. Peers can physically move or change their availability at any time. As such, protocols for redundancy must be explored to ensure client traffic does not fail when peer infrastructure changes.

Conclusion

The communication needs of displaced persons can be challenging to meet with current policies and technologies. In the above, we present a research agenda to generate advances in both domains to meet these needs. In the policy realm, recommendations were made for diffusing new approaches to spectrum management such that universal acceptance of more flexible spectrum use is achieved. As for new technologies, suggestions were made for research along two trajectories. The first involves networking technology that enables integration of diverse types of networks. The second views a bottom-up, easily deployed network operating in either a stand-alone mesh mode or passing data between networks. In combination, new approaches to policies and new technologies will benefit the displaced by providing reliable communications that in turn can offer a significant improvement to their well-being.

Acknowledgments

This work was funded through NSF Network Science and Engineering (NetSE) Award CNS-1064821 and NSF Catalyzing New International Collaborations (CNIC) Award IIA-1427873. We thank Dr. Nijad al-Najdawi and UNHCR staff for facilitating access to Za'atari.

Notes

1. https://info.internet.org/en/story/connectivity-lab.
2. https://www.google.com/loon.
3. http://www.gsma.com/newsroom/press-release/smartphones-account-two-thirds-worlds-mobile-market-2020.
4. https://www.icrc.org/eng/resources/documents/news-release/2012/jordan-news-2012-09-26.htm.
5. https://www.ushahidi.com.
6. https://freeswitch.org.
7. http://www.unhcr.org/4444afcb0.pdf.

References

Chowdhery, A., Chandra, R., Garnett, P., & Mitchell, P. (2012). Characterizing Spectrum Goodness for Dynamic Spectrum Access. *50th Allerton Conference on Communication, Control, and Computing*. Monticello, IL.

Congress, U. S. (2009). S.649—Radio Spectrum Inventory Act.

Ghaznavi, I., Heimerl, K., Muneer, U., Hamid, A., Ali, K., Parikh, T., et al. (2014). Rescue Base Station. *ACM DEV '14*. San Jose, CA.

Haider, B., Zafrullah, M., & Islam, M. (2009). Radio Frequency Optimization & QoS Evaluation in Operational GSM Network. *World Congress on Engineering and Computer Science*. San Francisco, CA.

Heimerl, K., & Brewer, E. (2010). *The Village Base Station*. San Francisco, CA: NSDR.

Heimerl, K., Hasan, S., Ali, K., Brewer, E., & Parikh, T. (2013). Local, Sustainable, Small-Scale Cellular Networks. *ICTD '13*. Cape Town, South Africa.

Iland, D., & Belding, E. (2014). EmergeNet: Robust, Rapidly Deployable Cellular Networks. *IEEE Communications Magazine, 52*, 74–80.

ITU. (2015). *Global Population Connectivity: ITU Facts and Figures*. ITU (International Telecommunications Union).

Johnson, D. L., Pejovic, V., Belding, E., & van Stam, G. (2012). VillageShare: Facilitating Content Generation and Sharing in Rural Networks. *ACM DEV 2012*. Atlanta, GA.

Kyriazakos, S., Karetsos, G., Gkroustiotis, E., Kechagias, C., & Fournogerakis, P. (2001). *Congestion Study and Resource Management in Cellular Networks of Present and Future Generations*. Barcelona, Spain: IST Mobile Summit.

Matinmikko, M., Palola, M., Mustonen, M., Rautio, T., Heikkilä, M., & Kippola, T., & Mäkeläinen, M., et al. (2015). Field Trial of Licensed Shared Access (LSA) with Enhanced LTE Resource Optimization and Incumbent Protection. *IEEE Dynamic Spectrum Access Networks (DySPAN)*. Stockholm, Sweden.

Nandagopal, T. (2016, April). *NSF Workshop on Spectrum Measurement Infrastructures*. Retrieved from http://www.cs.albany.edu/~mariya/nsf_smsmw.

Palola, M., Matinmikko, M., Prokkola, J., Mustonen, M., Heikkilä, M., & Kippola, T., & Heiska, K., et al. (2014). Live Field Trial of Licensed Shared Access (LSA) Concept Using LTE Network in 2.3 GHz Band. *IEEE Dynamic Spectrum Access Networks (DYSPAN)*. McLean, VA.

Pizzi, M. (2013, May 6). Logging on in Za'atari: Part I. Retrieved from SMEX: Social Media Exchange, http://www.smex.org/logging-on-in-zaatari-part-i/.

Rhizomatica. (2015, January 14). *So Much Going On!* Retrieved from Rhizomatica: Mobile Communications for All, https://rhizomatica.org/2015/01/14/so-much-going-on/.

Schmitt, P., Iland, D., Belding, E., Tomaszewski, B., Xu, Y., & Maitland, C. (2016a). Community-Level Access Divides: A Refugee Camp Case Study. *ICTD '16*. Ann Arbor, MI.

Schmitt, P., Iland, D., Zheleva, M., & Belding, E. (2016b). *HybridCell: Cellular Connectivity on the Fringes with Demand-Driven Local Cells*. San Francisco, CA: IEEE Infocom.

Schmitt, P., Raghavendra, R., & Belding, E. (2015). Internet Media Upload Caching for Poorly Connected Regions. *ACM DEV '15*. London, UK.

Sengupta, S. (2015, September 26). Mark Zuckerberg Announces Project to Connect Refugee Camps to the Internet. *New York Times*.

Vernon, A., Deriche, K., & Eisenhauer, S. (2016). *Connecting Refugees: How Internet and Mobile Connectivity can Improve Refugee Well-Being and Transform Humanitarian Action*. Geneva: UNHCR.

Vigil, M., Rantanen, M., & Belding, E. (2015). A First Look at Tribal Web Traffic. *WWW '15*. Florence, Italy.

Zheleva, M., Paul, A., Johnson, D. L., & Belding, E. (2013). Kwiizya: Local Cellular Network Services in Remote Areas. *Mobisys '13*. Taipei, Taiwan.

7 Information Systems and Technologies in Refugee Services

Carleen F. Maitland

The 21st century's forced migration crisis is taking place amid rapid technological change. From using web-based portals to match European hosts with refugees, to mobile money supported food distribution in Kenya, new ways of supporting refugees are emerging. The sustainability and scalability of these efforts will depend on the niche they fill, problem they solve, and their potential to replace or integrate with existing systems.

Aside from the occasional leapfrogging or disruptive innovation, ICTs tend to develop along technological trajectories. Their adoption and use are often influenced by their fit with organizational and end-user needs, as well as their place in industry power structures. Hence, the success of new systems requires knowledge of technological, organizational, and end-user contexts, and can also benefit from an understanding of their evolution.

To this end, this chapter describes a sample of information systems currently used to support refugees. These descriptions highlight several points and trends. The systems may be global, regional, or local. They may address problems specific to refugees, or instead target more general social welfare needs, such as child protection. Systemic innovation often is driven by new technologies, arising from broader societal trends such as security concerns. However, widespread adoption usually occurs only after economies of scale have driven down prices. Also, humanitarians accept technology is not always the solution, and are willing to retreat to less sophisticated approaches when failures occur. Finally, similar to new technology use in our personal lives, privacy and security concerns raised at the outset are either addressed, continue, or increased use creates a level of apathy driven by the technology's ubiquity.

These insights emerge from the chapter's organizational informatics lens. Embedded within the volume's broader sociotechnical systems frame, this subfield views systems development and use as emergent processes. As such, they shape and are shaped by dynamic technological, organizational, and social contexts (Kling, 2000; Lamb & Kling, 2003; Orlikowski, 1993; Orlikowski & Robey, 1991; Robey & Markus, 1988). Instead of seeking only singular, linear, and expected effects, organizational informatics research finds multiple as well as paradoxical effects of the same technologies across diverse organizations and organizational levels (Sawyer & Rosenbaum, 2000). These differences are often attributed to divergent organizational structures and forms of IT governance (Sambamurthy & Zmud, 1999). IT governance is shaped by a variety of factors, including structural traits of hierarchy and specialization, or the vertical and horizontal distribution of tasks within an organization (Pugh, Hickson, Hinings, & Turner, 1968; Williams & Karahanna, 2013).

This chapter also presents a research agenda that argues for critical analyses of current and future system impacts for all stakeholders (Kvasny & Richardson, 2006; Myers & Klein, 2011). Given the significant power differential between refugees and service providers, the research agenda particularly promotes independent analyses of the impacts of emerging technologies on the displaced. Such analyses, which require access to refugee populations, would complement those by humanitarian organizations, which until now are the primary source of knowledge of these impacts.

This chapter is organized as follows. First, general background on the management of humanitarian information systems is provided, differentiating it from the more familiar contexts of organizational informatics research. Next, a sample of information systems from the categories of "data production and capture," including mobile, biometrics, and registration systems, as well as "data management and transfer," including systems for coordinated case management and mobile cash, are discussed. Finally, the chapter concludes by articulating a future-oriented research agenda that is both scientifically interesting yet practically grounded.

Background

Compared with those in government social service agencies, humanitarian information systems management has similarities but important differences as well. In both sectors, IT management must contend with cultural and power divides between staff and beneficiaries, occasional resistance by staff to adopting new technologies, as well as competing, and usually higher paying, markets for IT staff, leading to frequent turnover. Despite these similarities, important differences exist in four key dimensions: beneficiaries, international operations, dynamic nature of interorganizational networks, and distance from funders.

First, compared to the typical clients of a governmental social service agency, the displaced are frequently more desperate. They are likely to have survived armed conflict or persecution by a repressive state, as well as lost their homes and possessions, and in the worst cases, loved ones. In their new community, they may lack the protection, assurances, and level of cultural similarity that accompanies citizenship. These differences have implications for the types of data humanitarian systems must handle, as well as beneficiaries' attitudes toward technology use. For example, the sharing of data on sexual and gender-based violence (SGBV), in cultures where securing such information can be even more critical, has created significant challenges.

Second, where humanitarian organizations operate internationally, their systems must contend with cultural and language differences, physical distance, and the need to deal with international diplomatic relations. Cultural differences may exist between the displaced, who provide the data, and humanitarian staff who enter, process, and manage it. With operations spanning time zones and contexts, systematic and standardized system deployments are a challenge. Yet, autonomy of local offices does create benefits for local control and configuration. However, it can leave local IT staff feeling undersupported. International operations also can require interaction with, and in some cases demurring to, the host country government on systems deployments. As noted in chapter 6, in the discussion of networks, humanitarian organizations may even face direct opposition from the host country government.

Third, humanitarian agencies' interorganizational networks are less stable. As humanitarian operations, budgets, and service providers change throughout the displacement lifecycle, circumstances and partnerships frequently change. These changes create unique challenges in deploying systems throughout the network of care providers (Maldonado, Maitland, & Tapia, 2009).

Finally, humanitarian organizations, whether international or local, have a different relationship with their funders. For state-run organizations, oversight and particularly accountability are a shared endeavor. For humanitarian organizations, funding comes from national governments and/or individuals who have no inherent basis for trust and also may be physically and culturally distant. This distance creates a greater emphasis on upward accountability (Christensen and Ebrahim, 2006), provided through monitoring and evaluation (M&E) units. Over the past decade, M&E operations have made greater use of IT, and increasingly favor biometric technologies.

In addition to biometrics, accountability innovations also include photography, bar codes, scanning, and RFID, which are used to improve efficiency and combat fraud. Fraud is a significant issue, and is carried out by refugees, as well as humanitarian and government staff, and those of firms involved in supply chains. While some fraud is detectable, it is not always clear-cut. Black markets may be signs of fraud, where refugees who cheat, receiving more than their share, can sell the items for cash. Black markets also may signal graft in the supply chain, with those responsible for distribution skimming goods and offering them for sale, sometimes even to refugees.

However, black markets may also reflect less nefarious forces. They may simply indicate beneficiaries prefer other goods, leading them to sell honestly obtained goods for cash to purchase preferred items. Problems can arise, however, if the preferred items are prohibited in the program, such as alcohol and cigarettes. Vouchers and cash can help align benefits to tastes, while limiting trade in prohibited items. Further, electronic vouchers in the form of swipe cards generate data that can be analyzed, similar to processes used by large grocery chains around the world. Analyses can detect fraud, improving effective supply and resource allocation, particularly for goods subject to spoilage. However, without proper authentication, vouchers can be traded or sold. Biometrics are being used to solve this authentication problem.

The use of digital vouchers, biometrics, and the like all require the involvement of refugees, many of whom are traumatized. Potentially having fled authoritarian regimes that use technology for tracking and control, some refugees may have an uneasy relationship with IT. Also critical is the heterogeneity of refugee populations and their dynamic nature. For example, as discussed later in this chapter, while a 2013 assessment of mobile phone use in the Dadaab refugee camp in Kenya showed very high penetration rates, a followup study in 2015 suggested new arrivals to the camp were much less likely to own a phone.

In the following sections, several systems are described, shedding light on their use contexts. The descriptions are developed though field research as well as interviews with staff from a variety of humanitarian organizations, including UNICEF, UNHCR, and WFP as well as their partners. We have also entered into two data sharing agreements with UNHCR, providing first-hand knowledge of data structures and processes for sharing. The sample of systems is not exhaustive, but balances widely deployed and critical systems that feed data throughout the sector with emerging systems only now starting to gain traction. The systems also reflect various motivations for sharing data, including the need to share data simply to carry out stand-alone humanitarian operations, as well as data sharing for coordinated case management through networks of service providers. These examples then form the basis for a discussion of research needed to inform future technological designs, organizational informatics theory, and the practice of supporting the displaced.

Data Production and Capture

In enterprise systems, the subsystems that capture data frequently serve as the foundation of the entire information architecture. Data capture processes, often the first step in developing a more techno-centric workflow in the transition from paper, may subsequently become integrated into a wide range of tasks. Data capture in refugee services is not an exception, with technologies such as mobile phones used across a wide range of operational functions. Here four data capture systems and technologies are presented. The discussion begins with mobile data capture, including a discussion of a broader system for centralized storage and management (Kobo). This is followed by another significant innovation in data capture—biometrics. Finally, the section concludes with a discussion of a refugee registration system, proGres.

Mobile Data Collection

As mobile devices have taken the world by storm, the humanitarian sector continues to find new and innovative uses. Even in the challenging tasks of rapid assessment and subsequent monitoring, these technologies have made significant changes, not only in the data collection process, but also in the types of data captured (location coordinates, photos, video). Mobile data collection systems consist of several components, including: (1) software to develop forms to be downloaded to and filled out on any mobile device, even without Internet connectivity; (2) the server to aggregate data from many different data collectors, also known as known as enumerators; and (3) the interface and analytic tools that enable simple analyses and visualizations.

One popular open source mobile data collection system used extensively in the humanitarian community is Open Data Kit (ODK), which was first developed in 2008 through collaboration between researchers at University of Washington and Google.org. Since its initial development, ODK has evolved into a platform, with a diverse community of independent developers and companies extending and tailoring its capabilities. One such effort is KoBo Toolbox, a joint effort of the Harvard Humanitarian Initiative and the UN Office for the Coordination of Humanitarian Affairs (OCHA). KoBo is designed specifically for the challenging contexts of humanitarian operations. It was first officially deployed in 2012, focusing on greater ease-of-use in creating complex data collection forms, collecting location, photo, video, and QR code data, as well as providing a limited range of basic analytic tools. KoBo also centralizes data storage, including questions and forms, a benefit for organizations with limited IT services.

In February 2016, UNHCR launched its own KoBo server to provide centralized data storage and analytic tools for its staff and partners. Similar to many KoBo platforms, the server is relatively open, allowing anyone to establish a user name and password, and thereby upload, store, and analyze data for their particular project. The server has already been used in item distribution operations by collecting data via barcode scanning. Whereas in the past, such efforts might result in locally stored data, the KoBo server enables a secure environment for managing data globally. The availability of a headquarters-sanctioned storage facility may in turn incentivize more digital data collection and can be used to support community-led programs such as our Community Asset Mapping project conducted in Za'atari camp in 2016 (Xu and Maitland, 2017).

In that project, we trained refugees to collect household data on assets, such as skills and education, via mobile phones. The refugees subsequently uploaded the data to UNHCR's KoBo server in Copenhagen, and then were able either to view automated analyses online or download the data for further analysis. The Kobo staff have released an API to facilitate development of custom analytic tools, fostering further use and hence centralization of data management in the organization. Similar systems targeting humanitarian organizations and others are being launched commercially.[1]

Mobile devices' geocoding capabilities, as well as their ease in annotating textual or numeric data with photos and videos, have extended their use. As discussed by Tomaszewski in chapter 8, location capture, or geocoding, which is relatively simple via mobile phones, has enabled the mapping of various types of data, from rapid assessments to monitoring and evaluation (M&E) efforts. As a result, assessments of everything from school programs to food distribution can generate map layers that when simultaneously presented provide a more complete picture, and spatial distribution, of activities.

Biometrics

Increasingly, across the humanitarian and migration communities, systems are being populated with biometric data. For migration systems, initial biometric use was driven by the security environment in the U.S. post 9/11, when the US-VISIT Program was introduced, driving growth of the biometric industry (IOM, 2005). The International Organization for Migration (IOM), an international NGO and UN affiliate, collects biometric data in the form of photos and fingerprints in its MIDAS system. The system was developed to help low-income countries better collect and manage data on all migrants entering and exiting their countries. Launched in 2009, as of 2016 it has been adopted in 19 countries across Africa and Latin America.

For UNHCR, biometric data collection became standard practice in registration in 2010 (UNHCR, 2013a). While identity and rights aspects of biometrics use with refugees are covered extensively by Kingston (this volume), here a greater emphasis is given to its evolution, integration with other systems, and, importantly, its role in accountability systems, an important driver of use.

The integration of biometrics into refugee registration was envisioned as early as 2001. Embedded in the larger Project Profile program, the separate biometrics effort, funded by the Dutch Government and developed by HSB Netherlands, focused on integrating digital fingerprinting technology into the registration database. Reviewing the project, UN auditors observed:

"While biometrics is seen as a very valuable tool for validating the registration process, it is a costly method to use. Further consideration of using the feature is therefore necessary and the Project Profile team should clarify in which situations the use of biometrics is recommended" (UN OIOS, 2006, p. 17).

The HSB system used a coded description of the fingerprint, rather than the fingerprint image itself, saving on storage and making the system more secure. However, concerns over security and privacy remained, and, as noted by auditors, the use of biometrics raised their visibility: "There are currently no clear guidelines for the privacy issues related to individual refugee data in electronic format, including data sharing within UNHCR, as well as with the host country, country of origin and country of (possible) resettlement. This issue becomes even more pertinent in light of the fingerprinting technology" (UN OIOS, 2006, p. 19).

The auditors further noted that while remote access to field data by headquarters was helpful to field staff as a form of remote IT support, it provided access to confidential data that was not yet governed by a formal (security) policy to guide and safeguard their use. The auditors recommended a policy be developed that cover not only access and use, but also data storage, archiving and sharing as well as validation of data received from third parties (UN OIOS, 2006). The data protection policy was finalized nearly 10 years later, in the Spring of 2015.

However, already in 2013, local implementations of biometrics had captured one million fingerprints and 500,000 iris records. Given this rapid growth, UNHCR sought to establish an agency-wide system with the following three goals: (1) to facilitate access to UNHCR services; (2) to prevent identity theft and identity substitution among the persons of concern (PoC); and (3) to reduce the risk of multiple registrations of the same person of concern under different identities. The system requirements specified integration with the registration database, and laid out a staggered implementation of fingerprint, iris, and facial scanning technologies (UNHCR, 2013b).

In addition to being ruggedized, the system was required to accommodate data collection from a diverse and sometimes traumatized population. UNHCR described the population as ranging in age from 0 to 60+ years old, as well as from people with degraded or poor quality fingerprints, missing fingers, persons with conditions affecting the iris (e.g., cataracts, persons with a missing eye, and persons with disfigured faces).

Security requirements were specified, both in data sharing as well as within the system itself. In data sharing, specifications included encryption and ISO 27001:2005 for Information Security Management Systems, as well as a requirement that system APIs/SDKs use the BioAPI 2.0 standard (or equivalent) for secure hardware connections. As for the system, security specifications included port management, use of firewalls, and use of strong passwords conforming to ISO 27001:2005.

In 2014, UNHCR proceeded with the design of a new Biometric Identity Management System (BIMS) and an associated Global Distribution Tool (GDT) (UNHCR, 2015). This system design was developed with insight gained from local implementations, and, as will be discussed below, now serves as a platform for integration with other systems and partners (UNHCR OIG, 2015). BIMS was developed in collaboration with Accenture, and includes fingerprinting and iris, as well as facial scan, technologies. After successful testing in Malawi in 2015, the system was initially deployed in Chad and Thailand (Accenture, 2015). In May 2017, the Kenya office of UNHCR reported use of BIMS in its registration system now integrated with the Government of Kenya. Interestingly, they also trained IOM personnel on BIMS in order to ensure refugees exiting the country had biometric-based identification.[2]

It should also be noted that UNHCR and its partners have also begun to use biometrics in their own human resource systems. For example, as shown in figure 7.1, employees in Za'atari refugee camp in Jordan, now use fingerprint technology as the basis for time sheets in their payroll system. This continued integration of biometrics into service systems, particularly across organizations, will be discussed further, following a discussion of proGres, in the section "Data Transfer, Sharing, and Management."

proGres

Registration, the initial step in refugee support, requires extensive data collection. Standardization in data structures and management practices during this initial and critical phase facilitates efficiencies within the registration agency, be that a host country government or UNHCR. More importantly, it also helps create uniform data quality, formats, and processes across the entire network of humanitarian organizations. In UNHCR, data quality is ultimately the responsibility of its Field Information and Coordination Support Section of its Division of Programme Support and Management.

Figure 7.1
Digital fingerprint time card. (Photo credit: Carleen F. Maitland.)

Where other parties, such as national governments, conduct registration, UNHCR provides guidance on registration data collection and information management.[3] In this role of centralized data collection, registration authorities must continue to evolve data collection practices, responding to trends such as broader recognition of gender categories, as well as relational roles beyond the traditional "married/single" dichotomy (ORAM, 2016).

For UNHCR and some national governments, registration is conducted using, and biometric data entered into, the proGres system. As registration can occur in remote locations, UNHCR designed the system for use both online and off and without power, on a laptop loaded with a stand-alone version.[4] The system has evolved significantly over the past 15 years, having first been developed in 1999 by Microsoft volunteers deployed during the Kosovo crisis. This genesis is emblematic of the influence of Corporate Social Responsibility programs. From there, in October 2001, UNHCR's

Executive Committee established "Project Profile" with the goal of standardizing and advancing registration practices, including further development of proGres (UN OIOS, 2006). As such, proGres became and continues to be UNHCR's main repository of refugee personal information. The system stores and manages information about family relations, refugee claims, cases related to refugee status determination, protection, resettlement, and other services. The system is continuously updated with information, such as the birth of a child, hospital visits, or new medical conditions. While biometric data including photos are now supported, the data entry process is usually conducted during a face-to-face interview, with data entered manually.

Over the course of the 2000s, the system's acceptance grew, and by the end of 2010, it was deployed in more than 250 locations in 82 countries.[5] By 2013, the system contained data on more than 4.8 million individuals, of which around 2.8 million records are active.[6] Over time, the system evolved and was based on a variety of Microsoft products, including a Dynamics CRM Enterprise Server, SharePoint for document management, BizTalk for Electronic Data Exchange (EDE), and Microsft SQL for database functionality (UNHCR & Microsoft, 2011).

proGres contains a variety of personally identifiable, as well as sensitive, information, where sensitive is defined relative to the context. For example, having fled one's country, indicated by simply registering for asylum, can be sensitive and potentially illegal, particularly in countries where exit visas are required or where the person fleeing is a military deserter. Even data often thought to be public, such as familial relations, can be sensitive for those moving from polygamous cultures or those seeking resettlement for extended family members. Registration data is also likely to include records of mental and physical health conditions, which can be highly sensitive. Potentially even more sensitive are data on sexual and gender-based violence (SGBV), of which there are higher occurrences in conflict environments.

In accordance with generally accepted principles of data ethics, the collection of sensitive data requires registration staff to obtain "informed consent." An example UNHCR refugee consent form reads: "I understand that in the following limited cases, UNHCR may need to share my basic bio-data (my name, the names of my dependents, my address and my telephone number in <country of asylum>) to ensure the provision of protection and assistance. In all cases, measures will be taken by UNHCR to prevent unauthorized dissemination of the information."[7] Gaining true informed consent in displacement is challenging due to the need to rapidly register, in some cases, large numbers of people, the registrant's lack of understanding of the system through which the data will move, and a registrant's (and sometimes also the registrar's) lack of understanding of data-related rights. However, this challenge is not limited to registration, as consent is required across the entirety of humanitarian data collection efforts (InterAction Protection Working Group, n.d.).

Safeguarding refugees' data is made more complex by the need to share data through organizational networks. This situation can create a conflict between protection and operational mandates, whereby sharing data increases risks, yet is often necessary to access specialized services. The conflict can become more acute when data sharing extends to host country governments. In some cases, governments taking over refugee status determination, as discussed by Ruffer (this volume), adopt proGres as their registration information system. The governments of Mozambique, Kenya, and Uganda, for instance, adopted proGres while taking over registration duties. The Kenyan case involved a 2016 MOU between the government and UNHCR allowing the Kenyan Department of Refugee Affairs (DRA) access to the UNHCR system, until the government's system was up and running.[8]

Together, these four systems of data production and capture (mobile devices, KoBo, biometrics, and proGres) are just several of many new sources of data used throughout humanitarian networks. With mobile and biometric technologies producing new forms of data, capture systems such as KoBo and proGres require continuous updates to accommodate and leverage these new assets. However, as discussed below, new data sources create even greater potential and challenges for transfer, sharing, and management.

Data Transfer, Sharing, and Management

Interorganizational data sharing is critical, and its success often depends on internal data management practices. In humanitarian organizations, the autonomy of field offices, allowing for localizing and customizing to local circumstances, can also make large scale processes and standard operating procedures, such as data quality management practices, challenging. Also, data are often controlled at local levels and program staff (e.g., shelter, food) are increasingly playing important roles alongside their IT counterparts in managing databases and systems. For example, within UNHCR, both the proGres and BIMS systems are managed jointly by IT and program staff.

These contextual forces shape the performance requirements of the interorganizational systems required for coordinated refugee care. In the following, three systems, ranging from global to local implementations, are discussed. The first two reflect efforts to improve service coordination, one with child protection and sensitive gender-based violence (GBV) data, and the other with critical refugee registration data. The third system encompasses digital cash and voucher systems, which are a recent and popular innovation.

Data Sharing with Primero

Refugee assistance is sometimes supported by global systems designed to address problems that transcend refugee contexts. In the area of child protection, UN agencies and their partners have developed the Primero system to foster interagency data sharing in a case management framework.[9] The system is managed by a steering committee consisting of UNICEF, IRC, Save the Children, UNFPA, and others. Primero is designed to conform to the data management standard articulated in the Minimum Standards for Child Protection in Humanitarian Action (UNICEF, 2015), and is an outgrowth of two earlier UN interagency initiatives.

The first, the Gender-Based Violence Information Management System (GBVIMS), is designed specifically to handle the highly sensitive data related to GBV. The GBVIMS initiative was originally launched in 2006 by UNOCHA, UNHCR, and the IRC, and is now in use in over 20 countries. It has been led by the GBVIMS Steering Committee consisting of UNFPA, UNICEF, UNHCR, IRC, and IMC.[10]

The second interagency system is the Child Protection Information Management System, an effort of UNICEF, Save the Children, and IRC, working together since 2005. By 2009, the system had been adopted by additional agencies, including Care, World Vision, and JRS, and was in use in 16 countries.[11] In 2016, UNHCR and Terre des Hommes Lausanne also joined the system's Steering Committee. The system's functionalities include case management, family tracing and reunification, data analysis, information sharing, data protection and confidentiality, as well as customizability to the local context.

Primero builds upon these and other systems, enabling their integrated use. The system is designed for deployment in a range of circumstances, with an emphasis on security and confidentiality. The latter is achieved with role-based access and granular security, ensuring only those who need to see data will have access to it via time-stamped, password-protected, and encrypted transactions (UNICEF, 2015).

Primero's underlying design philosophy is that data should be shareable and actionable, even highly sensitive GBV data.[12] Its design embraces beneficiary-based control, understanding that victims sometimes need to control who is aware of the abuse they have suffered. As a result, the system does not mandate sharing, allowing users to retain ownership of data and sharing according to information sharing protocols adapted to local contexts. Action is supported by built-in customizable reports, connecting data with decision-making on critical programs to support vulnerable children and survivors of GBV (UNICEF, 2015).

Since 2013, the initiative has been led by the UNICEF Child Protection Section at Headquarters on behalf of the interagency group. The software firm Quoin led Primero's development employing a user-centered and field-based design approach. Operations in Jordan, Kenya, and Somalia were consulted in the developing and testing of the prototype. The system was first implemented in Sierra Leone during the Ebola response.[13] A system demonstration conducted at the Innovation Marketplace, an event of the 2016 World Humanitarian Summit, generated further interest in the platform as well as discussion of the potential for integration with the aforementioned KoBo mobile data collection system (World Humanitarian Summit, 2016).

Data Sharing with RAIS

As discussed above, the proGres system, which holds each refugee's unique identity number as well as demographic information, can serve as a powerful platform for data sharing with service providers, greatly improving coordination. However, concerns over data privacy, security, and integrity often foreclose this option.

UNHCR and its regional partners in the Middle East overcame this problem by creating the Refugee Assistance Information System (RAIS), a web-based platform making some data fields in proGres more accessible, and at the same time helping UNHCR and the entire service network gain access to data of implementing partners. Unlike Primero, RAIS is currently a regional effort. While sharing data among organizations does generate risks, the system is seen as enhancing security over what was the previous standard practice—data being shared in spreadsheets via email.[14]

Spurred by the influx of Iraqi refugees in Jordan (Davis & Taylor, 2012), more than 30 UNHCR implementing partners adopted RAIS in 2010. As a modular web application, RAIS helps keep refugee information up to date and secure. For service providers, it helps plan targeted assistance to persons of concern, prevents duplication of efforts, provides efficient reporting tools, and reduces fraud and abuse of assistance services. In addition to its coordination benefits, RAIS provides insight into the complex nature of service provision and enabled the regional Data Analysis Group to move from time-constrained to continuous data analyses. In turn, this helped the network of providers to better target services during a time at which they were facing budget cuts (UNHCR 2010).

As described by a research team of academics working with UNHCR: "The RAIS actively receives health information on registered Iraqi asylum seekers and refugees in Jordan from more than 30 partnering organizations at 100 centers, including nongovernmental organizations (NGOs), primary health-care clinics, hospitals, pharmacies and government-sponsored medical centers" (Mateen et al., 2012, p. 2).[15] They also note the types of data captured and stored in RAIS during a healthcare visit, including patient name, date of birth, date of visit, sex, diagnosis, type of care (acute versus chronic), type of evaluation (inpatient or outpatient), use of a medical procedure, and referral to a medical specialty.

As the Syrian catastrophe followed closely on the heels of the Iraqi crisis, UNHCR in Jordan decided to use the system for this much larger population. That growth spurred the decision to expand the system's use to Lebanon and Egypt. The upgrade was designed to be generally applicable, rather than specified to any one country or any single urban location. The upgrade expanded the system from two to ten modules, including a smart search function for refugee profiles, an improved referral system to facilitate communication between partners, a reports module, mobile data collection to facilitate entry during home visits, and a service to manage the appeal to donors for refugee assistance. With the addition of these modules, use of RAIS has grown, including by more than 200 implementing and operational partners in these countries.

Digital Cash and Voucher-Based Systems

During the last decade, greater recognition of the need for choice, control, and dignity of beneficiaries spurred a movement toward cash-based assistance. Yet, this approach can reduce control over how funds are used, a critical element for accountability. Digital cash and digital vouchers, as opposed to physical cash, can help retain some elements of control. In the following discussion the coevolution of digital cash and voucher services, together with biometrics systems, is presented. The three separate systems, implemented in Kenya, Rwanda, and Jordan, all involve food distribution programs operated jointly by the World Food Program (WFP) and UNHCR. Unlike the more widely deployed systems discussed above, each has its own unique elements, as approaches continue to evolve.

The first system was a trial implemented at the Kakuma and Dadaab refugee camps in Kenya. The collaboration began not as a digital voucher or cash program, but as a biometric system to augment a barcode-based ration card system. The project's goal was to address the widely known problem of "double dipping" at a camp now more than 25 years old. To enable UNHCR to share biometric data stored in proGres, the two agencies signed a global MOU in 2011. Based on this agreement, in 2013 the Dadaab project was initiated with the following goals:

> (i) the protection and confidentiality of refugee data; (ii) the regular update and maintenance of the registration system (proGres) before each food distribution cycle; (iii) the access to the UNHCR network (proGres) at the final delivery points (FDP) only by staff with valid authorization; (iv) the development of procedures for registering alternate food collectors (AFC) and (v) the handling of litigation [addressing authentication problems] by trained and experienced staff. (UNHCR OIG, 2015)

The eventual system was comprised of servers, laptops, barcode readers, and fingerprint scanners, and a variety of security controls. At the heart of the system is a network of laptops, a proGres/fingerprint terminal server loaded with databases required to verify fingerprints, photos, and other data for each refugee, and a series of connected laptops used to scan ration card barcodes, beneficiary fingerprints, and collect other data (UNHCR OIG, 2015). In addition to physically securing equipment, data were encrypted and standard wireless LAN security tools employed. Data integrity is facilitated through recording of communication between the WFP litigation counters, where problems with authentication are resolved, and the database server at UNHCR. The system also undergoes regular backups, and reliability is ensured through the option to deploy a mobile server, particularly for the case of power failures (UNHCR OIG, 2015).

A joint assessment by the Offices of Inspector Generals, which are internal oversight bodies in both UNCHR and WFP, concluded the system was highly effective, generating several benefits in the areas of finances and overall coordination. In particular, the OIG reported:

> The biometrics project implemented in Dadaab and Kakuma led to a substantial decrease in estimated refugee populations. Since its implementation on 1 October 2013, the number of people receiving food assistance has dropped significantly, attributed largely to a reduction in the number of ration cards being used fraudulently. The cost reductions for WFP resulting from the implementation of the biometrics for food distribution in Kenya are above USD 1.5 million *per month* [emphasis added]. (UNHCR OIG, 2015, p. 3)

And further, the report stated:

> According to a survey conducted by WFP in November 2014, 77 percent of refugees polled were satisfied with how the new controls had been implemented, and 60 percent responded that the new procedures made food distribution faster and more orderly. (UNHCR OIG, 2015, p. 26)

The project also generated longer-term benefits. The project's cost savings generated an estimated 1300% return-on-investment over five years. This enhanced the overall program's credibility. Also, donors who were previously hesitant to contribute to food aid, precisely due to the fraud, were now interested. The project also benefitted coordination between the two agencies. As noted in the oversight report:

> The effective collaboration of UNHCR and WFP during the planning, design and implementation phases of the system contributed to enhanced mutual appreciation of each other's outlooks, concerns and challenges. WFP Kenya acquired a better understanding of UNHCR's refugee protection mandate and related issues; while UNHCR Kenya learned about WFP's food distribution and pipeline challenges. (UNHCR OIG, 2015, p. 9)

In 2015, the two agencies teamed up again to launch a digital voucher trial. The program, called Bamba Chakula, had the goal of expanding food choices, as at the time rations focused on durable staples, rather than fresh options. For those refugees with additional sources of income, fresh options had always been available in the camp through private vendors. If successful, this program would make those options available to all by allocating a certain percentage of the monetary value of food rations to such purchases.

To support refugee choice and organizational control, the program recruited a select group of the camp's established food vendors. The vendors were vetted for, among other criteria, quality of food, consistency of operations, and willingness to receive digital payments. The choice of the payment system was obvious, given the popularity of Safaricom's M-PESA service among the camp's refugees. The Bamba Chakula system required refugees seeking the mobile payment to first pass the biometric clearance, according to the aforementioned process. Only then would the funds be released to their M-PESA account, which can then be used in transactions with approved food vendors. The scale of the program is very large, involving over 100,000 households and more than 300 vendors. The plan was to start slowly, allocating just 10% of the ration amount and then increasing, as the system is stabilized, up to 30%.

In 2016, a nearby settlement established for newly arriving South Sudanese refugees, has adopted the Bamba Chukula program in a more extensive form. Nearly all, as opposed to 10–30%, of the food ration value is distributed through the system. In this experimental settlement, called Kalobeyei, refugees are meant to fully integrate with the host community. Instead of establishing the settlement with barriers and then later working to break them down, in this settlement the two groups work and live side-by-side. The Bamba Chakula program provides employment opportunities for both groups, filling positions as the program's vendors.

Although yet to be confirmed by independent evaluation, the Bamba Chakula program appears to be a success. In contrast, results of a similar program in Rwanda, at least at the outset, were less so. This second program also involved WFP and UNHCR, but in this project implementation partners included World Vision Rwanda, the Bank of Kigali (BoK), Visa, and Airtel, a regional mobile communication provider. The program, begun in December 2013, was rolled out in three camps, and, similar to Bamba Chakula, enabled refugees to transact for food from vetted local merchants. However, there were several differences. First, not all refugees had mobile handsets, so they needed to be distributed. Lacking a mechanism for providing widely accepted mobile payments, as compared to M-PESA in Kenya, the system instead used SMS-based payments between the refugees and vendors. For the mobile payments partners, Visa and Airtel, this project was part of a broader push to develop these services in the region.

World Vision's role in the partnership was that of a "help desk" for refugees struggling to use the system. It turned out there were many challenges, particularly related to the phone, but also due to the numerous partners and different perspectives on the project. World Vision reported problems such as poor network connectivity, stolen handsets, and forgotten pin numbers. In addition to end-user technical problems, the system also faced challenges in timely account updates on behalf of the bank, which made it difficult to know the account balances.

With the contract with Bank of Kigali, now I&M Bank, due to expire, WFP and UNHCR decided to switch tactics, away from mobile phones, to an approach that more fully aligned with their goals for financial inclusion. As a result, the mobile phone-based system was replaced with a smart card. The new partnership involved WFP and UNHCR, together with Equity Bank. In this program, Equity Bank enrolled refugees in its banking services, similar to the process for opening a regular bank account. Refugees were provided a card, with transactions authenticated via PIN or with digital fingerprints. In this system, Equity Bank collects and stores the digital fingerprints itself, as compared to obtaining them through data sharing with UNHCR. Authentication is managed by the point-of-sale terminals used by Equity Bank agents and merchants, which are equipped with both a PIN pad and a single finger/thumbprint scanner.

This program differs from its mobile predecessor in two important ways. First, the smart card establishes a relationship between the refugee and the full service bank, which was not available through the previous Rwandan program nor is directly facilitated by the M-PESA-based system in Kenya. The smart card more directly fulfills goals for financial inclusion. Second, while refugees no longer have support for mobile phones and SIM cards, they are also relieved of the pressure to keep the phone charged (a challenge where electricity is scarce) and have the biometric option if they forget their PIN.

The third and final digital cash/voucher system is one deployed in the Za'atari Syrian refugee camp in Jordan. The initial system used a digital voucher, in the form of a card (see figure 7.2). The program was similar to the one in Kenya in that it did not enable refugees to receive cash, only to spend the balance in their "wallet" with approved vendors. However, unlike Kenya and Rwanda, in Za'atari they did not vet local merchants.

Figure 7.2
Digital voucher in the form of a card.

Instead refugees have to purchase goods through two centralized grocery stores operated by WFP. However, similar to both the Kenyan and Rwandan programs, the digital voucher serves as a platform with multiple wallets, enabling other humanitarian organizations to distribute digital benefits as well.

While reasonably successful, the WFP Za'atari project continued to struggle with fraud. In response, in 2016, they launched a system, their global first in its ability to use iris scan technology to both authenticate and debit an account. Using the aforementioned global MOU, WFP accesses the iris scans held in UNHCR's proGres database. With the ability to authenticate via iris scan, it quickly became clear it was possible to do away with the voucher cards altogether. The system, which relies on the biometrics technology, known as EyePay, has been described by the WFP country director, Mageed Yahia, as "a milestone in the evolution of our food assistance program."[16]

According to WFP, the iris-based system is more efficient, enforces accountability (meaning reducing fraud), and makes grocery shopping easier and more secure for the refugees. Issues such as lost cards and forgotten pin codes are eliminated by the new system. Benefits also accrue to the host country, Jordan, as not only are Jordanian staff becoming trained on these advanced systems, the EyePay technology was developed by the Jordanian firm IrisGuard.

IrisGuard has extensive experience in biometrics. As early as 2001, they deployed biometric-based international border systems in the UAE and Jordan, followed in 2008 by deployment of the world's first iris-based ATM machines. Similar to the UNHCR/WFP system, customers were relieved of their bank cards and pin codes, with the system resulting in hundreds of millions of dollars of cardless cash transactions. In 2013, IrisGuard extended the use of the technology from mere cash withdrawals to a full payment platform called EyePay. The platform, designed for use in retail stores and online shopping, is the technology deployed by WFP.[17]

WFP's deployment requires coordination between UNHCR's biometric data, and partners Jordan Ahli Bank and Middle East Payment Systems (MEPS). The system checks the shopper's iris details captured at the grocery store and confirms it with UNHCR's database. It then checks the refugee's balance through the MEPS financial gateway, creates the debit, and generates a receipt for the refugee. WFP plans to deploy the system in all Syrian refugee camps in Jordan and may expand beyond the camp setting as well.[18]

The development of these systems, and particularly insights gleaned from the OIG review of the Kenyan system, provide important lessons as well as cautionary tales. First, in the OIG's report, it appears a difference of opinion arose with the OIG preferring a more thorough use of automatic notifications when authentication failed, to which the UNHCR/WFP Kenya staff were opposed.[19] At its core, this conflict reflects different views on the use of technology, with the OIG wanting consistent use and perhaps putting greater trust in the technology, while the local office wanted to allow for exceptions. In systems design, it is important to recognize these differences are likely to arise and establish mediation processes a priori.

Second, organizations should be prepared to assess a system's broader impacts, such as the local food supply. While conditions for Kenyans purchasing food at markets *outside* the camp could be argued as "beyond the scope" of the current project's measurable impacts, a negative impact could have significant implications for UNHCR's relations with the local community and in turn the government of Kenya. Indeed, the OIG noted that the WFP had observed the biometrics project coincided with a considerable decrease in the amount of food aid for sale in the local markets, but fortunately no rise in prices.

Third, planning, implementation, and reporting should more clearly and consistently document resistance to these projects, by staff and refugees alike. It was not until the end of the OIG report that the following explanation appeared: "The low rate of awareness (20%) [of the project] observed in Dadaab camps was mostly due to the fact that refugee leaders were vehemently opposed to the introduction of the new controls, and had actively sabotaged the communication campaign." (UNHCR OIG, 2015, p. 32). It can be assumed the "vehement opposition" was related to the dismantling of the system generating extra income through fraud; however, the reasons for the opposition are not explained. Where systems confront fraud this opposition should be planned for and made explicit. Also, opposition should be highlighted and analyzed.

Research Directions and Conclusions

The continuing evolution of organizational systems in displacement relief services provides many valuable lessons for practice as well as insight for theories related to organizational informatics. The information systems described above, including proGres, KoBo, RAIS, Primero, and the cash-based systems, reflect trends and challenges faced by the sector. They also suggest at least five compelling issues to be addressed by organizational informatics scholarship.

First, an important area for scholarship is the issue of conflict as reflected in these systems deployments, and, more importantly, understanding its management and implications. Conflict can arise in establishing the boundaries of acceptable use of information technology, or in resistance to new systems by refugees, staff, or both. Research is needed that clarifies how these conflicts intersect with issues such as: (1) governance and autonomy within and between organizations; (2) power differentials between agencies and beneficiaries; and (3) relations between agencies and local communities as well as host governments.

Second, greater understanding of how innovation is managed through humanitarian networks is needed. The systems described here vary from top-down initiatives (Primero) to those germinating from the ground up (Dadaab digital voucher). Systematic investigations of innovation processes require access to multiple organizations and ideally multiple field sites. Dedicated innovation units within organizations can help provide such access, since the findings are surely to further support their missions. Of particular interest are studies of the ways in which technologies disseminate through networks, especially involving non-UN (IGOs and local NGOs) implementing partners. It would also be helpful to know how technologies disseminate from one organizational function to another. For example, from food distribution, which of the many operations, including sanitation, health, and education, is most likely to integrate digital cash into their operations? How deep into the networks of service providers will the effects of these innovations ripple?

Third, given the wide range of emerging technologies that might be used in refugee service provision, it is important to understand their differential impacts. Comparative analyses of the use of different technologies across the same functions (e.g., bar codes scanning versus iris scans), as well as across different functions, will provide important insights for organizational informatics on fundamental technical effects. What benefits will newer technologies bring and to whom? For example, facial scans may provide little improvement in benefits vis-à-vis iris scans in authentication and identification, but may provide significant benefit to reuniting families. Organizational informatics analyses can help identify these benefits, taking into account the governance and authority structures needed to reap benefits.

Fourth, as alluded to above, research on the implications of technological trends will be critical for strategic planning for the sector. Understanding the implications for innovations in machine learning, 3D video, big data, and virtual reality for the full range of refugee services can help establish priorities and even potentially shape the direction of technological developments in ways that benefit the sector. Further, such analyses can help promote development of more robust and longer-lasting policies in areas such as data privacy and security, and establish more robust decision-making criteria.

Finally, and most importantly, there is a need for analyses focused on the perspectives of the displaced. As technologies evolve and are further integrated into beneficiary services, refugees are likely to encounter technology in their homes, in services, in receiving information, and in mediating their relationships with service providers, host country nationals, host country governments, and even one another. Important research questions, intersecting social and organizational informatics, include: (1) How is the experience of displacement overall affected by the broad range of information and communication technologies? (2) What impact do information systems in the provision of service have on refugees' lives, and perceptions of and experiences with technology? (3) From refugees' perspectives, which technologies used in service provision have the greatest, least, best, and worst implications for the quality of service received or simply quality of life?

More specifically, and directly related to technologies currently in use, research might examine: (1) What role do refugees play in managing their own biometric data? (2) What are the long term consequences of this data capture? (3) What level of transparency exists over data access and sharing? This last question ties into broader social concerns over managing privacy in ever more digitally connected lives, involving privacy statements and policies and user/consumer activism in detecting violations of data sharing or data being sold.

While UNHCR typically has the best interests of refugees in mind, they face many pressures, including resource and staffing constraints, the need to meet accountability requirements posed by donors, the avoidance of conflict with host country governments, and the pressure to implement programs that benefit the majority of refugees rather than simply a few. These pressures may result in decisions biased toward the organization's needs over those they serve. In these circumstances, independent analyses are critical. Academics and independent NGOs, such as Refugees International, can contribute important insights. Humanitarian organizations should embrace the critique and opportunity for self-reflection spurred by such investigations as critical to the process of innovation.

As technologies and the systems through which they are deployed continue to evolve, systematic inquiries can generate knowledge of value to researchers and practitioners alike. This knowledge can provide the basis for improved practice in the short term and advanced theories and improved technologies in the long term.

Notes

1. See Ona, a Kenyan-based social enterprise offering centralized data storage and management.

2. http://www.unhcr.org/ke/wp-content/uploads/sites/2/2017/06/Kenya-Operation-Factsheet-May-2017.pdf.

3. http://data.unhcr.org/imtoolkit/chapters/view/registration-in-emergencies/lang:eng.

4. Remote locations may be the ideal scenario for registration, because, while a refugee may wait or be unable to register until they reach a capital city or urban center, the preferred method is to register the displaced as soon as they cross the border.

5. Microsoft, 2011, download.microsoft.com/.../b/3/.../unhcrprogressolutionoverview.pdf.

6. An individual's details are marked as "closed" in proGres only when the individual ceases to be of concern to UNHCR.

7. https://docs.google.com/document/d/1C7Fm8NFP4s3ESnK00rdcA-BiKQZe3eeUnwtHrZJzOSY/edit.

8. http://reliefweb.int/report/kenya/unhcr-kenya-factsheet-april-2016.

9. http://www.primero.org/resources.

10. http://www.gbvims.com/

11. http://www.cpaor.net/sites/default/files/cp/Evaluation-IA-CP-IMS-Final-ENG_0.pdf.

12. Personal communication, Robert MacTavish, UNICEF, February 3, 2016.

13. https://www.quoininc.com/projects/unicef-primero.

14. http://www.unhcr.org/innovation/labs_post/rais-refugee-assistance-information-system

15. http://www.who.int/bulletin/volumes/90/6/11-097048/en.

16. http://www.huffingtonpost.com/entry/syrian-refugees-eye-scan_us_56c4c9bae4b08ffac1275343.

17. http://www.euromoneyconferences.com/Jordan-Speaker-Details.html?id=8580

18. https://www.wfp.org/news/news-release/world-food-programme-uses-innovative-iris-scan-technology-provide-food-assistance-.

19. Specifically, OIG described the lack of a warning system for what might appear to be an attempt to access the system twice as "a conscious decision from UNHCR and WFP management in order not to restrict freedom of refugees as a matter of principle." UNHCR Kenya's response noted that first a failed biometric match may indicate many other problems than just potential fraud, and second that other mechanisms exist to thwart fraud. OIG further commented that the proposed warning system "is not perceived by the Inspectors as a restriction to refugees' freedom and right to collect food." This difference lies at the heart of the missions of both UNHCR and WFP, but here we see two entities internal to both organizations differing on how best to use technology to carry out that mission.

References

Accenture LLP. (2015). UNHCR: Innovative Identity Management System Uses Biometrics to Better Serve Refugees. Retrieved from https://www.accenture.com/t20161026T063323__w__/us-en/_acnmedia/Accenture/Conversion-Assets/DotCom/Documents/Global/PDF/Dualpub_15/Accenture-Unhcr-Innovative-Identity-Management-System.pdf.

Davis, R., & Taylor, A. (2012). *Urban Refugees in Amman, Jordan*. Washington, DC: Institute for the Study of International Migration.

InterAction Protection Working Group. (n.d.). *Data Collection in Humanitarian Response: A Guide for Incorporating Protection*. Washington, DC.

Kling, R. (2000). Learning About Information Technologies and Social Change: The Contribution of Social Informatics. *Information Society, 16*(3), 217–232. doi:10.1080/01972240050133661.

Kvasny, L., & Richardson, H. (2006). Critical Research in Information Systems: Looking Forward, Looking Back. *Information Technology & People, 19*(3), 196–202. doi:10.1108/09593840610689813.

Lamb, R., & Kling, R. (2003). Reconceptualizing Users as Social Actors. *Management Information Systems Quarterly, 27*(2), 197–236.

Maldonado, E. A., Maitland, C. F., & Tapia, A. H. (2009). Collaborative Systems Development in Disaster Relief: The Impact of Multi-level Governance. *Information Systems Frontiers, 12*, 9–27. doi:10.1007/s10796-009-9166-z.

Mateen, F. J., Carone, M., Al-Saedy, H., Nyce, S., Ghosn, J., Mutuerandu, Ti., et al. (2012). Medical Conditions Among Iraqi Refugees in Jordan: Data from the United Nations Refugee Assistance Information System. *Bulletin of the World Health Organization, 90*, 444–451. Retrieved from doi:10.2471/BLT.11.097048.

Myers, M. D., & Klein, H. K. (2011). A Set of Principles for Conducting Critical Research in Information Systems. *Management Information Systems Quarterly, 35*(1), 17–36.

Ngamassi, L., Maldonado, E., Zhao, K., Robinson, H., Maitland, C., & Tapia, A. H. (2011). Exploring Barriers to Coordination Between Humanitarian NGOs: A Comparative Case Study of Two NGOs Information Technology Coordination Bodies. *International Journal of Information Systems and Social Change (IJISSC), Special Issue on IS/IT in Nonprofits, 2*(2), 1–25.

ORAM. (2016). Incorporating Sexual and Gender Minorities Into Refugee and Asylum Intake and Registration Systems. Retrieved from http://oramrefugee.org/wp-content/uploads/2016/05/Registeration-Forms-Memo-English-1.pdf.

Orlikowski, W. J. (1993). CASE Tools as Organizational Change: Investigating Incremental and Radical Changes in Systems Development. *Management Information Systems Quarterly, 17*(3), 309–340.

Orlikowski, W. J., & Robey, D. (1991). Information Technology and the Structuring of Organizations. *Information Systems Research, 2*(2), 143–169.

Pugh, D. S., Hickson, D. J., Hinings, C. R., & Turner, C. (1968). Dimensions of Organization Structure. *Administrative Science Quarterly, 13*(1), 65–105. doi:10.2307/2391262.

Robey, D., & Markus, M. L. (1988). Information Technology and Organizational Change: Causal Structure in Theory and Research. *Management Science, 34*(5), 583–598. doi:10.2307/2632080?ref=no-x-route:122d6dd6c03f934528825f5a26dba853.

Sambamurthy, V., & Zmud, R. W. (1999). Arrangements for Information Technology Governance: A Theory of Multiple Contingencies. *Management Information Systems Quarterly, 23*(2), 261–290. doi:10.2307/249754.

Sawyer, S., & Rosenbaum, H. (2000). Social Informatics in the Information Sciences: Current Activities and Emerging Directions. *Informing Science, 3*(2), 88–95.

Tafere, M., Katkiwirize, S., Kamau, E. N., & Nsabimana, J. (2014). Mobile Money Systems for Humanitarian Delivery: World Vision Cash Transfer Project in Gihembe Refugee Camp, Rwanda. In L. R. Vazquez & I. Will (Eds.), *Communications Technology and Humanitarian Delivery: Challenges and Opportunities for Security Risk Management* (pp. 42–44). European Interagency Security Forum (EISF).

UN OIOS. (2006). Project Profile. Retrieved from http://download.cabledrum.net/wikileaks_archive/file/un-oios/OIOS-20060518-01.pdf.

UNHCR. (2013a). Biometric Identity Management System. Retrieved from http://www.unhcr.org/550c304c9.pdf.

UNHCR. (2013b). Request for Proposal: No. RFP/2012/507 For the Provision of a Biometric Identity Management System. Retrieved from http://www.unhcr.org/admin/sts/50c85dd69/request-proposal-rfp2012507-provision-biometric-identity-management-system.html.

UNHCR. (2015). Global Strategic Priorities Progress Report.

UNHCR, & Microsoft. (2011). proGres Supporting Refugee Registration and Camp Management. Microsoft Corporation.

UNHCR OIG. (2015). Joint Inspection of the Biometrics Identification System for Food Distribution in Kenya. Retrieved from http://documents.wfp.org/stellent/groups/public/documents/reports/wfp277842.pdf.

UNICEF. (2015). Primero Project Brief.

Williams, C. K., & Karahanna, E. (2013). Causal Explanation in the Coordinating Process: A Critical Realist Case Study of Federated IT Governance Structures. *Management Information Systems Quarterly*, *37*(3), 933–964.

World Humanitarian Summit. (2016). Primero "Debut" at the World Humanitarian Summit.

Xu, Y., & Maitland, C. (2017) Mobilizing Assets: Data-Driven Community Development with Refugees. In Proceedings of ICTD '17, Lahore, Pakistan, November 16–19, 2017. doi: 10.1145/3136560.3136579.

8 Geographic Information Systems (GIS) and Displacement

Brian Tomaszewski

Displacement is inherently spatial. The spatial aspect of displacement uniquely positions Geographic Information Systems (GIS) for documenting, representing, analyzing, curating, and understanding displacement situations within the complex body of information systems used in service providers networks for planning, decision-making, and operations. This chapter first explores two areas related to GIS and displacement—advances in GIS for challenging environments, such as refugee situations, and the theoretical foundations of GIS and displacement. Following and building upon these discussions, further directions for GIS and displacement research are presented.

GIS for Challenging Environments

GIS for challenging environments has seen significant progress in the past twenty years. Most notably in this regard is the application of GIS technology in the allied field of disaster management to support situation awareness. For example, the early 1990s saw the first calls for incorporation of GIS in all phases of disaster management (Coppock, 1995; Johnson, 1992). Over the intervening years, advances in general computing hardware (PCs, GPS, smartphones), commercial GIS software, location-based services such as Google Maps, public and private sector satellite data, geo-enabled social media, map-based crowdsourcing, and documented case studies of the value GIS provides for disaster management have strengthened GIS's position as a vital disaster management resource (Tomaszewski, 2015; Tomaszewski, Judex, Szarzynski, Radestock, & Wirkus, 2015a).

Refugee Camp Observation and Planning

Of particular note is the use of GIS for (a) camp planning (United Nations High Commision for Refugees, 2015) (Westrope & Poisson, 2013), and (b) earth observation (EO) data to analyze refugee/IDP camp population numbers, the built environment of a camp (such as building types in a camp) and overall growth and change of a camp over time (Bjorgo, 2000; Kemper & Heinzel, 2014). The uses of GIS for camp observation, planning, and maintenance are numerous. For example, staff in the Azraq camp in Jordan recently developed a GIS-based shelter allocation program (UNHCR Innovation, 2016). The use of GIS as a tool for camp planning and management has long existed within UNHCR and continues to be promoted (UNHCR, n.d.).

Field-Based Mapping and Refugee Participatory Mapping

As compared to the relatively longer term and centrally managed operations of camp planning, GIS has also been adopted in time-sensitive and more operationally diffuse processes. In the context of natural disasters, rapid mapping deployment has now become a standard, particularly in international natural disasters. On-the-ground mapping activities have also seen advancement in terms of utilizing digital mapping tools such as Open Data Kit[1] for registration and camp monitoring (REACH, n.d.-a, n.d.-b) and Survey123 (Esri, 2015).

Additionally, specialist NGOs like MapAction[2] have a well-established track record for rapidly deploying to disaster zones to establish spatial data collection and development of mapping products that can keep pace with the rapid and uncertain nature of disaster response (Tomaszewski, 2015). Groups like MapAction have also matured to the point that they are now part of the established UN System for dealing with actual disasters and often work with groups such as the United Nations Disaster Assessment and Coordination (UNDAC) organization (Tomaszewski, Judex, et al., 2015a).

GIS technology in displacement, particularly for field-based mapping, also has a long tradition and has seen notable advancement. An example of this advancement is observable in the work of the REACH initiative[3] at the Za'atari Syrian refugee camp in Jordan. Za'atari, in the context of Jordan, is particularly notable for the sheer volume of Syrian refugees that entered the camp between 2012 and 2013. Use of maps and geographical information were particularly essential to managing the rapid growth of the camp. As the camp quickly grew from an unplanned and temporary location for refugees into a planned settlement, this information was crucial to employing concepts from urban planning to create a more orderly environment. As an example of the positive outcomes of the enhanced usability of GIS technologies, refugees themselves played an important role in developing maps for Za'atari.[4]

A particularly interesting development in the use of GIS for challenging displacement environments is the transition from rapid, response mapping (like the aforementioned examples) to mapping for longer term planning and recovery. In this context, the "challenge" is not so much a lack of electrical power or datasets (although these remain recurring issues), as is typical in a disaster response, but rather, the capacity to use previously collected geographic information and GIS software's analytic capability to inform ongoing longer term planning in a refugee camp.

For example, the Dadaab refugee camp in Kenya, first established in 1998, is the largest refugee camp in the world. In this context, GIS is being used for examining the camp's overall impacts on the surrounding region, in areas such as access to drinking water and general environmental impact (Beaudou, Cambrézy, & Souris, 1999; UN Institute for Training and Research or UNOSAT, 2011).

The Za'atari refugee camp is also transitioning into a phase of using GIS for longer term planning and management of the camp and refugee community. In this context, GIS is used for continuing maintenance of electrical networks, water and sewers, and many tasks often associated with city government and planning offices (figure 8.1).

Despite the aforementioned example, GIS use by refugees is far less prevalent than use by staff. The obvious barriers to such practice stem from a lack of basic resources, including computers, electricity, and technical skills. However, it could be argued that camp residents are perhaps in the best position to understand spatial dynamics of the camps in which they live.

Again using Za'atari as an example, this settlement can be considered an urban refugee camp, meaning it has access to electricity, computers, and other capacities necessary for using GIS. Based on these resources, Za'atari has the potential, over time, to serve as an exemplar of combining GIS education with technical capacity building, to allow refugees to manage,

Figure 8.1
The Za'atari refugee camp in Jordan is seeing development of long-term infrastructure such as electrical networks. (Photo credit: Brian Tomaszewski, 2016.)

create, and analyze spatial data related to issues relevant to their community. For example, issues might include identifying locations of community assets where information is transferred (such as a market), providing inputs to longer term planning processes (such as electrical grid updates), identifying locations of potential vulnerability to hazards (such as winter storms), and allowing refugees to map what they consider "safe areas" and neighborhoods within a camp (figures 8.2a and 8.2b).

Empowering refugees to use GIS themselves for decision-making presents a key opportunity to give voice to the displaced and non-UN/INGO actors. For example, in 2017, Za'atari camp began a novel program called Refugee GIS, or RefuGIS, that is putting GIS capacity in the hands of refugees themselves (Tomaszewski et al, 2017). The RefuGIS project is already showing results in terms of creating voice and empowerment for refugees as evidenced in this quote from Yusef Hamad, a Syrian refugee who has been living in Za'atari camp for four years and is a member of the RefuGIS team: "I quit my well-paid job to be in this [the RefuGIS] project; I look at it as my ticket out of Za'atari camp."

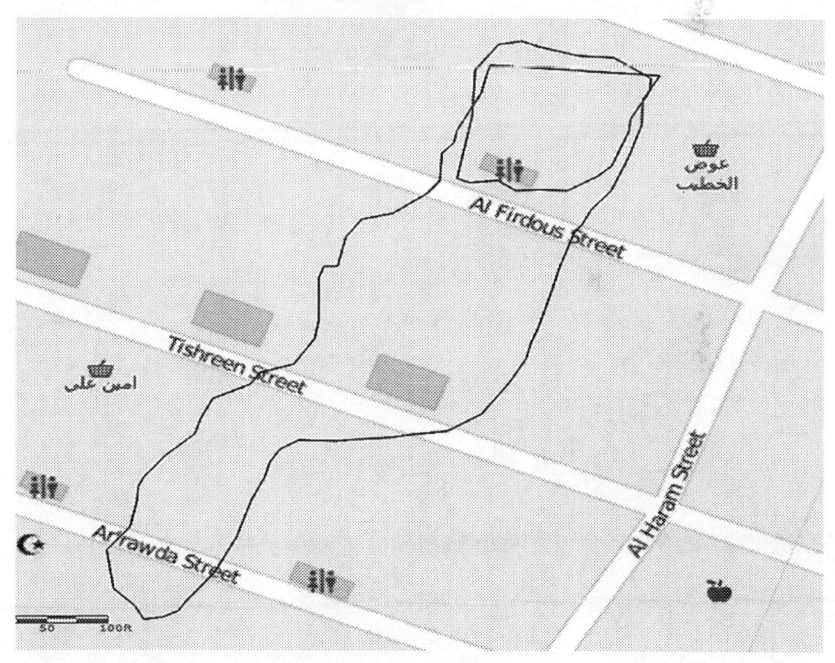

Figures 8.2a and 8.2b

Syrian refugee using tablet computers to map "safe areas" and neighborhoods within Za'atari camp. "Safe areas" and neighborhoods are unique, personal spatial phenomena that do not necessarily conform to established streets, blocks, and districts that are used to organize the camp for administrative purposes.

Complementing the sophisticated mapping and GIS-based analyses that have and are being conducted in camps such as Za'atari and Dadaab are efforts to create basic reference data and maps for refugee camps after the initial crisis has subsided. For example, in 2016 a group from the Rochester Institute of Technology (RIT), led by the author of this chapter, worked with UNHCR and refugees themselves in Rwanda to map the Kigeme Congolese refugee camp (figure 8.3).

The Kigeme camp-mapping mission is also an example of how low-cost/free, easily accessible mapping technologies can be used for rapid field data collection. For example, within five days, my group of undergraduate students from RIT were able to map the entire camp infrastructure, including administrative boundaries, WASH stations, schools, and refugee businesses (Tomaszewski et al., 2016b) (figure 8.4). I was particularly impressed with how well the students were able to use low-cost/free, easily accessible mapping technologies to rapidly collect data. For example, most of my students had little technical background or class work in using GIS. However, they were all very comfortable with using smartphones in general, so found it relatively easy to using smartphone-based mapping tools. The approach of

Figure 8.3
A Congolese refugee mapping the Kigeme refugee camp in Rwanda. (Photo credit: Brian Tomaszewski, 2016.)

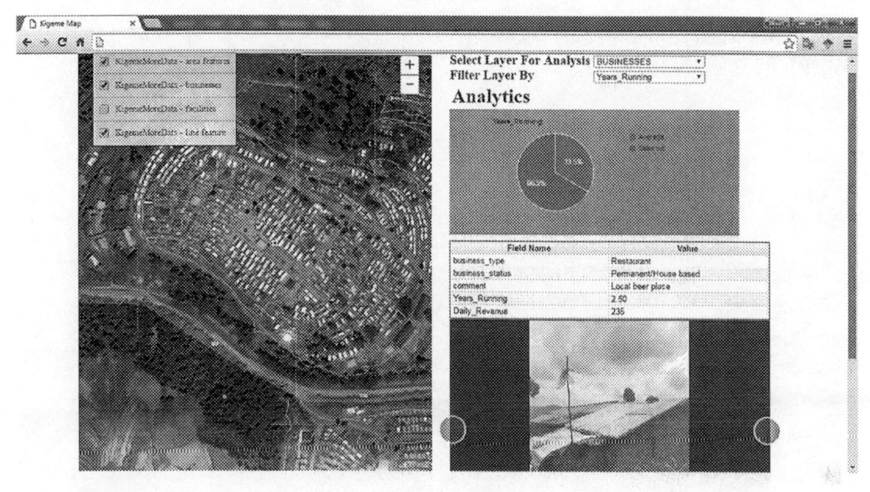

Figure 8.4

Kigeme refugee camp, Rwanda. The map on the left shows samples of data collected by undergraduate college students from the US and Congolese refugee camp residents. Items on the right are analytic tools being developed to support UNHCR operations. For example, the picture shown on the bottom right is of an illegal alcohol selling business. Such activities are of particular interest to protection officers working in the camp who are trying to prevent children from accessing alcohol. By using a map, camp officials are provided a better picture of the camp's spatial dynamics for a wide variety of sectors such as livelihood, protection, WASH, and others.

using smartphone-based mapping tools like Collector for ArcGIS (http://www.esri.com/products/collector-for-arcgis) or Open Data Kit (ODK) Collect (https://opendatakit.org/use/collect/) is a recommendation I can make for practitioners as well. For example, if there is a need to do rapid and low- to no-cost mapping, such tools could be utilized to support mapping teams with little background in mapping. All that is needed is one person who is experienced with the tools and their setup to then instruct the non-experienced people on how to conduct the actual mapping tasks. This was the approach I used my with students that led to great success in mapping a refugee camp with over 20,000 people in just four business days.

In addition to collecting basic reference data for the camp, the Kigeme camp-mapping mission was also important in that it brought awareness of GIS, mapping, and spatial thinking to camp management officials, many of whom were not familiar with GIS and its representational and analytic power (figures 8.5 and 8.6).

Figure 8.5
Participants from UNHCR and other Rwandan refugee stakeholder groups learning about GIS software at a workshop in Rwanda. The map image shows datasets collected during the Kigeme mapping mission. (Photo credit: Brian Tomaszewski, 2016.)

Figure 8.6
In the same workshop, participants were shown how to use modern, easy-to-use, smartphone-based mapping tools. In this figure, a UNHCR camp official is learning how to map a polygon feature. (Photo credit: Brian Tomaszewski.)

GIS and Displacement Theoretical Foundations

Analyses of the relationship between GIS and displacement can benefit from two theoretical foundations: critical cartography and refugee situation awareness.

Critical Cartography

Critical cartography considers the process of creating maps and final map representations as expressions of power and political processes (Crampton & Krygier, 2005). A critical cartography perspective is useful for considering how GIS and maps created by GIS tools contribute to political processes of refugees, host countries, and broader political mechanisms. As an example, Palestinian refugee cartographies have been seen as "threats" by Jewish Israelis (Weaver, 2014). Issues related to digital spatial data access and sharing (or lack thereof) have also been documented in this region (Maitland, 2013). Kelly (2015) has investigated critical cartographies related to mapping of the experiences with crossing borders and experiences working in camps, specifically addressing the question: "How can the cartographic portrayal of Syrian peoples' border crossings be improved to better represent their experiences?" Rygiel (2012) investigated how camp spaces reproduce and reform power structures. Madden and Ross (2009) investigated the mapping of personal stories and areas using Google Earth, an idea sometimes referred to as "maptivism," or using maps to advocate, visualize, and make sense of complex issues (Kreutz, 2009).

To the best of our knowledge, little work has been conducted that critically examines refugee camp map products. Such practice can occasionally be found in the field of disaster management, particularly in after-action reports that review how a particular agency performed during a disaster.[5] Given the differences in the processes, persons, and agencies involved in general disaster management versus the specific domain of displacement, more research is needed on theoretical issues of social power, representation, and political structure of interest to critical cartographers as an important contribution to the basic research on displacement. Additionally, further research should be conducted on the utility of map products generated by organizations such as UNHCR and their NGO partners and the value of those maps in solving problems and improving the lives of refugees.

Refugee Situation Awareness

The second theoretical lens is refugee situation awareness, which reflects the idea "how the various actors that serve refugees and the refugees themselves can maintain awareness of factors related to their own personal wellbeing as well as the situation in their home country—or refugees and people that support them" (Tomaszewski, Mohamad, & Hamad, 2015b, p. 3). The idea is based on the more general situation awareness concept, which is a theoretical construct used to predict and measure awareness levels in dynamic, evolving environments, such as vehicle operation, military situations, and disaster management (Endsley, 1995; Endsley, 2000). In terms of GIS and displacement, "refugee situation awareness" as a theoretical construct primarily has been utilized (although perhaps not using the extract term) for empirical, observation-based research like the aforementioned research on using EO data and spatial analysis techniques to analyze refugee camp environments. Less research attention has been paid to situation awareness of refugees themselves.

GIS and Displacement Research Questions and Policy Implications

Combining the advances in technology with the aforementioned theoretical foundations of critical cartography and situation awareness generates questions in four general areas, including: spatial analyses of refugee geographies; multiscale refugee situation awareness and spatial thinking through education; the spatial dynamics of refugee resilience; and the spatial dimensions of the effects of cash-for-work programs. Each is presented in turn below.

Research Questions

1. Preliminary insights from my work in Za'atari camp and observations of the many maps that can be found throughout the camp shed light on the problem of how refugee cartographies and geographies represent a re-creation, creation, modification, or existing regional spatial patterns of displacement? Refugee geographies are typically mapped at (1) country scale, in terms of origin and destination country; and (2) the general layout of refugee camps and population estimates. Little to no research has been conducted that maps, at the camp scale, (a) where refugees in a camp

originate from in a host country, and (b) compares camp-based human geographies to broader regional patterns and host country displacement geographies. Additionally, given that GIS is a surveillance technology, such investigations must take precautions to protect refugee privacy (Sui, 2011).

2. My experiences in Jordan and Rwanda (where I have also conducted spatial thinking education research) suggest it is important to develop further knowledge as to: What role can GIS and maps play for multiscale refugee situation awareness and spatial thinking? It is well known that generally when maps and the ability to create maps and digital spatial data and products are given to people, their overall spatial thinking and situation awareness will improve (Lee & Bednarz, 2012; Tomaszewski, Vodacek, Parody, & Holt, 2015c). However, much work remains on how such interventions can have impact on the displaced. For example, use of smartphone-based maps has been well documented as the navigation tool of Syrian refugees migrating across Europe. However, little attention has been paid to how camp-based refugees can specifically benefit from mapping and GIS. With education being a strong focus of many displaced communities, more attention should be paid to incorporating spatial thinking and GIS education as part of a general ICT curriculum that is given to refugees. This may allow refugees themselves to have better agency over their own lives in camps and potentially develop livelihood strategies related to mapping and GIS as this is a growing global industry (Tomaszewski, Mohamad, et al., 2015b).

3. Displacement situations and natural disaster risk often coincide. Refugee camps are often placed in undesirable locations in terms of land use. In my experiences in Za'atari camp Jordan, I have seen weather extremes that go from intense winter storms to flood-inducing rain to unbearable heat in the dessert. In Rwanda, which is a very hilly country, refugee camps are often located on steep hillsides susceptible to landslides and flooding. From these personal observations, further cases of protracted displacement situations, and the growing academic interest in natural disaster resilience, it is important to understand: What are the spatial dynamics of displacement in relation to factors such as refugee resilience, and how can these factors be indexed and quantified? For example, refugee resilience is a top priority for the Syrian Regional Crisis.[6] Spatial indexes of refugee disaster

resilience currently do not exist—let alone at the community-level scale. Most resilience indictors proposed in the natural disaster literature are at the county-level scale (Cutter, Burton, & Emrich, 2010). Additionally, refugee resilience has seen some work in terms of planning relocations upon arrival in a host country due to stressors such as climate change that can exacerbate displacement (Weerasinghe, 2014). However, much work remains to develop a refugee disaster resilience index to quantify and evaluate how disaster resilience is manifested in displaced populations that face unique circumstances. Development of refugee disaster resilience indicators and their subsequent mapping and visualization using GIS can be used to support decision-making and analysis for refugee service provision similar to disaster risk maps, as Li et al. (2013) of Lanzhou University, China notes. Community-level scale field mapping of refugee resilience indicators (such as community capacities) is one starting point for spatially indexing and quantifying refugee disaster resilience with GIS.

4. In terms of research to support development, advocacy, and opportunities for refugees themselves, as I have seen through the RefuGIS project in Jordan and livelihood policy shifts in Rwanda, future research should examine spatial dimensions of how cash-for-work programs have modified livelihoods within the camp. For example, based on insights derived from the aforementioned mapping activities, it has been speculated that the diversity of businesses in Kigeme camp has grown since UNHCR introduced a cash-based system for food delivery as opposed to the traditional in-kind donations.

Policy Implications

Policy implications for the use of GIS and displacement contexts are primarily related to sharing of GIS data collected by governments and the international community (i.e., the UN and their implementation partners). In most cases, formal policies for such arrangements do not exist. More often, it is the case that Memoranda of Understanding (MoUs) exist where governments make offers to share data collected by outside agencies working in the country. However, often these arrangements can be hampered by a lack of existing data policies in general, or a lack of capacity to share spatial data from technological and organizational perspectives. It is important to note as well that issues related to policies promoting sharing of governmental and international community GIS data are broader than the displacement context. Governments may not have data-sharing policies or lack the capacity to share data. For example, in Jordan, development of a National Spatial Data Infrastructure (NSDI) is still in a relatively early stage due to the typical issues (such as technical capacity, human resources and training, and funding) that are needed to implement NSDIs (IDRC/ESRI Canada, 2011).

Conversely, the UN system has its own policies regarding data sharing. For example, UNHCR established an interagency information-sharing portal for coordination of the Syrian Regional Refugee Response (figure 8.7).

In terms of data-sharing policy, GIS, and mapping, Za'atari camp has been extensively mapped with the spatial data for the camp being stored in Open Street Map,[7] thus making camp data available to wide a range of humanitarian actors with capacity to use modern digital mapping tools. Despite promising advances like collection of refugee camp data stored in Open Street Map, there is still much that can be done to formalize (and enforce) policies that promote data sharing among the UN System, NGOs, host country governments, and refugees themselves. A first step in this regard is recognition of the need for data collection, procurement, curation, and sharing when a displacement response occurs. The following quote from

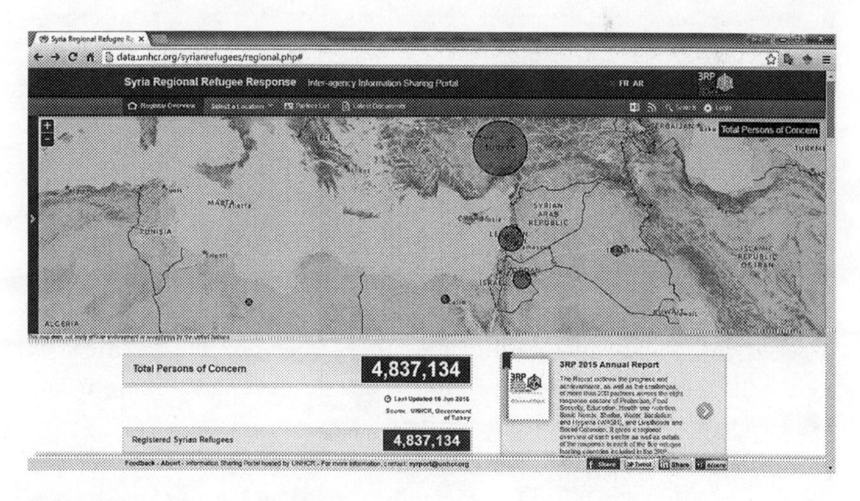

Figure 8.7
The UNHCR Syrian Regional Refugee Response information sharing portal. Source: http://data.unhcr.org/syrianrefugees/regional.php.

Lóránt Czárán[8] (United Nations Office for Outer Space Affairs or UN-OOSA) captures the sentiment of this idea in the context of a disaster response, but is equally applicable in displacement response. I have used italics to emphasize key points Czárán makes for data policies in displacement:

> There are also others, mechanisms such as the Central Emergency Response Fund (CERF) that various countries put together so the UN can faster react to any major disaster situations. I would love to see provisions for that fund to concretely support acquisition of satellite or geospatial data, just the same way as today the same fund supports acquisition of food or shelter or other materials. Have provisions for rapid new data acquisition, data management as well in the context of these funding mechanisms that are designed to react quickly when a disaster happens, because that's where we are still lagging way behind. *For some reason, investment in data is never a priority. Sometimes senior management expects it to just appear for free. It might, but maybe in a week or maybe in two. And by that time it's too late for a number of reasons. And I'm convinced that if such funding mechanisms could be used for the quick data procurement just as they are used today for other elements, we might make much better progress in terms of using geospatial data for supporting the disaster management phases and everything in that context.* (Quoted in Tomaszewski, 2015, p. 182)

Summary and Conclusions

Geographic Information Systems (GIS) have made great progress in the past twenty years for spatial data management, curation, analysis, and decision-making in disaster and displacement contexts. As demonstrated in this chapter, GIS has progressed from initial use as a refugee camp observation and planning tool to deeper use for real-time data collection during rapidly evolving crises and use of GIS technology by refugees themselves. As displacement situations become more protracted, and GIS technology and data more available to UNHCR and other refugee service stakeholders, GIS will become a more vital part of longer-term strategic planning for both camp-based and urban refugees. Additionally, GIS presents unique educational and livelihood opportunities for refugees themselves due to the need for a skilled workforce that can collect, manage, and analyze refugee data like seen in the refuGIS project in Jordan.

Theoretical foundations of relevance to GIS and displacement include, but are not limited to, critical cartography, which can inform how the process of mapping and map products themselves shape discourse around refugee and displacement situations. On a more operational level, the concept of refugee situation awareness may potentially inform how refugees can use maps and digital mapping tools for developing insight into their displacement situation. In a trans-national migration situation, this could involve use of maps on smartphones for navigating in a new city or finding service locations. In a camp setting, situation awareness could involve understanding day-to-day life in a camp, such as locations of meetings or community outreach events, or changes in camp morphology (Tomaszewski, Tibbetts, Hamad, & Al-Najdawi, 2016a). In any case, refugee situation awareness operates on multiple spatio-temporal scales, since refugees will have the need to understand what is happening in their countries of origin .

Combining the technological and theoretical perspectives, the chapter then presented select areas for future GIS and displacement research that range from theoretical advancement to operational research for refugee support. These included: (a) spatial analysis of refugee geographies; (b) multiscale refugee situation awareness and spatial thinking through educational intervention; (c) the spatial dynamics of refugee resilience; and (d) the spatial dimensions of cash-for-work programs and the impacts of such programs on refugee livelihoods. There is, of course, much more research that can be done at the intersection of GIS and displacement. It is hoped that readers of this chapter who are new to spatial analysis and visual representation of displacement will use these topics as starting points for other types of spatial analyses and investigations into the broader role of GIS in displacement.

The chapter concluded with perspectives on policy implications for GIS and displacement in terms of data sharing. Data-sharing policy implications for GIS and displacement have much in common with those in other crisis contexts such as natural disasters. However, the key difference with displacement is the potential for longer, protracted situations that require sustained commitment and sharing from a wider variety of stakeholders (i.e., host governments, UNHCR, NGOs, the refugees themselves) who need to share and coordinate data.

GIS in crisis situations is not new. The same can be said for displacement. However, significant efforts are needed to further the incorporation of spatial perspectives and methodologies into displacement research, as well as to further adoption of GIS technology and ideas into displacement service provisioning. Ideally, making such advancements can improve, if not save, the lives of those faced with displacement.

Notes

1. https://opendatakit.org.

2. http://www.mapaction.org.

3. http://www.reach-initiative.org.

4. http://www.reach-initiative.org/informing-humanitarian-action-with-gis-in-al-zaatari-camp-2.

5. See US Department of Homeland Security (DHS) (2006) for a review of FEMA after hurricane Katrina, and the value and limitations maps in geospatial data played.

6. http://www.3rpsyriacrisis.org.

7. http://wiki.openstreetmap.org/wiki/Refugee_Camp_Mapping#Case_Study:_Refugee_Camp_in_Jordan.

8. The views expressed herein are those of the author(s) and do not necessarily reflect the views of the United Nations.

References

Beaudou, A., Cambrézy, L., & Souris, M. (1999). *Environment, Cartography, Demography and Geographical Information System in the Refugee Camps Dadaab, Kakuma–Kenya: UNHCR–IRD*. ORSTOM.

Bjorgo, E. (2000). Using Very High Spatial Resolution Multispectral Satellite Sensor Imagery to Monitor Refugee Camps. *International Journal of Remote Sensing, 21*(3), 611–616.

Coppock, J. T. (1995). GIS and Natural Hazards: An Overview from a GIS Perspective. In A. Carrara & F. Guzzetti (Eds.), *Geographical Information Systems in Assessing Natural Hazards* (pp. 21–34). New York: Springer.

Crampton, J. W., & Krygier, J. (2005). An Introduction to Critical Cartography. *ACME: An International E-Journal for Critical Geographies, 4*(1), 11–33.

Cutter, S. L., Burton, C. G., & Emrich, C. T. (2010). Disaster Resilience Indicators for Benchmarking Baseline Conditions. *Journal of Homeland Security and Emergency Management, 7*(1).

Endsley, M. R. (1995). Toward a Theory of Situation Awareness in Dynamic Systems. *Human Factors, 37*(1), 32–64.

Endsley, M. R. (2000). Theoretical Underpinnings of Situation Awareness: A Critical Review. In M. R. Endsley & M. Garland (Eds.), *Situation Awareness Analysis and Measurement* (pp. 3–32). Boca Raton, FL: CRC Press.

Esri. (2015). Survey123 Helps Direct Relief Provide Medical Care to Refugees. Retrieved from http://www.esri.com/esri-news/arcnews/fall15articles/survey123-helps-direct-relief-provide-medical-care-to-refugees.

IDRC/ESRI Canada. (2011). Feasibility Study for a National Spatial Data Infrastructure in Jordan. Washington, DC: infoDev/World Bank.

Jing, L., Liu, X. M., & Gang, L. (2013). Public Participatory Risk Mapping for Community-Based Urban Disaster Mitigation. *Applied Mechanics and Materials, 380,* 4609–4613.

Johnson, G. O. (1992). GIS Applications in Emergency Management. *URISA Journal, 4,* 66–72.

Kelly, M. (2015). Mapping Syrian Refugee Border Crossings: A Critical, Feminist Analysis. Retrieved from http://mappingborders.github.io/index.html.

Kemper, T., & Heinzel, J. (2014). Mapping and Monitoring of Refugees and Internally Displaced People Using EO Data. In Q. Weng (Ed.), *Global Urban Monitoring and Assessment through Earth Observation* (p. 195). Boca Raton, FL: CRC Press.

Kreutz, C. (2009). Maptivism: Maps for Activism, Transparency and Engagement. Retrieved from http://www.crisscrossed.net/2009/09/14/maptivism-maps-for-activism-transparency-and-engagement.

Lee, J., & Bednarz, R. (2012). Components of Spatial Thinking: Evidence from a Spatial Thinking Ability Test. *Journal of Geography, 111*(1), 15–26.

Jing, L., Liu, X. M. & Gang, L. (2013). Public Participatory Risk Mapping for Community-Based Urban Disaster Mitigation. *Applied Mechanics and Materials 380,* 4609–4613.

Madden, M., & Ross, A. (2009). Genocide and GIScience: Integrating Personal Narratives and Geographic Information Science to Study Human Rights. *Professional Geographer, 61*(4), 508–526.

Maitland, C. (2013). *Maps, Politics and Data Sharing: A Palestinian Puzzle.* Paper presented at the Proceedings of the Sixth International Conference on Information and Communications Technologies and Development: Notes–Volume 2, Cape Town, South Africa.

REACH. (n.d.-a). Informing Humanitarian Action with GIS in Al-Za'atari Camp. Retrieved from http://www.reach-initiative.org/informing-humanitarian-action-with-gis-in-al-zaatari-camp-2.

REACH. (n.d.-b). Rapid Mapping of Storm Damage in Za'atari Camp, Jordan. Retrieved from http://www.reach-initiative.org/rapid-mapping-of-storm-damage-in-zaatari-camp-jordan.

Rygiel, K. (2012). Politicizing Camps: Forging Transgressive Citizenships In and Through Transit. *Citizenship Studies, 16*(5–6), 807–825.

Sui, D. (2011). Legal and Ethical Issues of Using Geospatial Technologies in Society. In T. L. Nyerges, H. Couclelis, & R. McMaster (Eds.), *The SAGE Handbook of GIS and Society*. Thousand Oaks, CA: SAGE Publications Ltd.

Tomaszewski, B. (2015). *Geographic Information Systems for Disaster Management*. Boca Raton, FL: Taylor and Francis (distributed through CRC Press).

Tomaszewski, B., Judex, M., Szarzynski, J., Radestock, C., & Wirkus, L. (2015a). Geographic Information Systems for Disaster Response: A Review. *Journal of Homeland Security and Emergency Management—Special issue on Information and Communication Technology (ICT) and Crisis. Disaster, and Catastrophe Management, 12*(3), 571–602.

Tomaszewski, B., Mohamad, F. A., & Hamad, Y. (2015b). *Refugee Situation Awareness: Camps and Beyond*. Paper presented at the Humanitarian Technology: Science, Systems and Global Impact 2015, HumTech2015, Boston, MA.

Tomaszewski, B., Tibbetts, S., Hamad, Y., & Al-Najdawi, N. (2016a). *Infrastructure Evolution Analysis via Remote Sensing in an Urban Refugee Camp—Evidence from Za'atari*. Paper presented at the Humanitarian Technology: Science, Systems and Global Impact 2016, HumTech2016, Boston, MA.

Tomaszewski, B., Tomaszewski, K. J., Forbes, J., Hartz, J., Heintz, A., & Womble, N., & Bhatia, V., et al. (2016b). *Displacement and Visual Analytics: Case Studies and Applications*. Paper presented at the SpatialVA workshop @ GIScience 2016, Montréal, Canada.

Tomaszewski, B., Vodacek, A., Parody, R., & Holt, N. (2015c). Spatial Thinking Ability Assessment in Rwandan Secondary Schools: Baseline Results. *Journal of Geography, 114*(2), 39–48.

Tomaszewski, B., Martin, J.-L., Omondi, I., Al-Najdawi, N., Tedmori, S. & Hamad, Y. (2017). Using Geographic Information Systems (GIS) in Za'atari Refugee Camp, Jordan for Refugee Community Information Management and Mobilization: The RefuGIS Project. In Cunningham, P., ed. IEEE Global Humanitarian Technology Conference (GHTC), San Jose, CA.

UN Institute for Training and Research (UNOSAT). (2011). Overview Map of Refugee Camps in Dadaab, Garissa, N. Eastern, Kenya Retrieved from http://reliefweb.int/map/kenya/overview-map-refugee-camps-dadaab-garissa-n-eastern-kenya.

UNHCR. (n.d.). Resources—The UN Refugee Agency. Retrieved from http://unhcragencies.weebly.com/resources.html.

UNHR/Innovation (2016). Using GIS Technology to Map Shelter Allocation in Azraq Refugee Camp. Retrieved from http://innovation.unhcr.org/using-gis-technology-map-shelter-allocation-azraq-refugee-camp.

United Nations High Commission for Refugees. (2015). Maps. Retrieved from http://www.unhcr.org/pages/49c3646c4ca.html.

US Department of Homeland Security (DHS). (2006). A Performance Review of FEMA's Disaster Management Activities in Response to Hurricane Katrina. In Department of Homeland Security: Office of Inspector General (Ed.). Retrieved from https://www.oig.dhs.gov/assets/Mgmt/OIG_06-32_Mar06.pdf.

Weaver, A. E. (2014). *Mapping Exile and Return: Palestinian Dispossession and a Political Theology of Shared Future*. Minneapolis, MN: Fortress Press.

Weerasinghe, S. (2014). *Planned Relocation, Disasters and Climate Change: Consolidating Good Practices and Preparing for the Future*. Washington, DC: Brookings Institution.

Westrope, C., & Poisson, E. (2013). Using GIS as a Planning and Coordination Tool in Refugee Camps in South Sudan. Retrieved from http://odihpn.org/magazine/using-gis-as-a-planning-and-coordination-tool-in-refugee-camps-in-south-sudan.

9 Data Analytics and Displacement: Using Big Data to Forecast Mass Movements of People

Susan F. Martin and Lisa Singh

In recent years, the global population of people forcibly displaced by conflict and persecution has reached levels unprecedented since World War II. Acute natural hazards also lead to large-scale movement of people, some temporarily and others permanently. Since 2008, the number of disaster-displaced averaged 21.5 million each year (IDMC, 2016). In many cases, both human-made and natural factors precipitate large-scale displacement, as witnessed by recurrent famines in Somalia caused by the confluence of drought, conflict, and political instability that impede access to food relief. This chapter identifies novel big data sources, methodologies, and challenges that need to be addressed in order to develop more robust, timely, and reliable evidence-based systems for detecting and forecasting forced migration in the context of humanitarian crises.

Patterns of forced migration in anticipation of, during, and following conflict and acute natural hazards are notoriously difficult to predict. What appear to be very similar pre-existing stressors and triggering events and processes can result in significantly different levels, forms, and destinations of displacement. The warning signs of displacement or significant changes in the nature of movement are often present but difficult to piece together in a coherent fashion. However, given the unprecedented levels of forced displacement, an urgent need exists to develop an evidence-based early warning system that can enable governments and international organizations to formulate contingency plans, establish appropriate policies, and pre-position shelter, food, medicines, and other supplies in areas that are likely to receive large numbers of refugees and displaced persons. Because much forced migration is unexpected, communities can be overwhelmed by refugees and displaced persons if they have insufficient warning. Even relatively wealthy countries may fall victim. The recent massive movement of Syrian, Afghan, Iraqi, and other asylum seekers into Greece, with the hope of moving onwards to the rest of Europe, is a clear example of such chaos.

Progress has been made in recent decades in establishing early warning systems to alert the international community as well as national and local actors of impending humanitarian crises. For example, tsunami and famine early warning systems monitor and analyze data relevant in anticipating acute and slow onset crises, respectively, relying on scientific, technological, economic, social, and other indicators (See FEWS Net and NOAA National Tsunami Warning Center).[1] Predicting crises in other domains, such as conflict and violence, has proven more difficult, but organizations such as the International Crisis Group put out regular alerts of worsening conditions,[2] and ACLED, the Armed Conflict Location and Event Dataset, codes the actions of rebels, governments, and militias within unstable states, specifying the exact location and date of battle events, transfers of military control, headquarter establishment, civilian violence, and rioting.[3]

Our research into these systems has identified a number of problems that must be solved to improve the effectiveness of early warning systems, particularly as they apply to displacement: (1) identifying and collecting masses of timely, reliable data on the complex factors that affect flight; (2) developing analytic capability to discover indicators of movement—specifically, leading indicators that displacement will occur rather than trailing indicators that confirm that movement has already taken place; (3) instituting mechanisms to allow policy makers and practitioners to test out scenarios to determine if actions will have positive or negative consequences in averting displacement or providing better assistance and protection; and (4) building the political will to act on the warnings. New technologies and analytic tools make it more likely that the first three problems can be tackled. The fourth problem is, of course, more difficult to solve, but more effective early warning tools might challenge political leaders to act, at least in implementing more timely emergency relief operations.

We focus our discussion on the following directions of research that are necessary to harness potential benefits of big data for anticipating patterns of forced migration: the development and validation of a theoretical model of forced migration that captures the complexity and dynamism of the phenomenon; the identification and collection of relevant data related to the complex factors that affect flight; the need for methods that take disparate forms of data with varying degrees of reliability and completeness and extract meaningful, timely evidence of movement; and the development of analytic tools that enable policy makers and practitioners to test different scenarios to respond to forecasted movements. Through this discussion, we describe initial efforts to use newspaper and social media data to begin generating direct and indirect indicators of movement. While big data has some important limitations, including the ratio of noise to signal that can distort the accuracy of forecasts and potential biases that exist because of incomplete data, these diverse data can be used by researchers to capture fragments of human behavior at large scale, in real time; these data are not always available using traditional approaches. It is the combination of the traditional survey data, available administrative data, and new structured data values extracted from big data sources that make early warning systems for forced migration plausible. While obstacles still exist for early warning tools in this area, the growing number of available sources and the advances in technologies make this an area where significant progress can be made over the next decade.

Development of a Theoretical Model of Forced Migration

Our work on early warning focuses on displacement in the context of humanitarian crises, that is, any situation in which there is a widespread threat to life, physical safety, health, or basic subsistence that is beyond the coping capacity of individuals and the communities in which they reside (Martin et al., 2014). The global population of people forcibly displaced by conflict and persecution exceeded 66 million in 2016. Between 2015 and 2017, violence and conflict caused new, large-scale displacement, particularly in the Middle East, Africa, and South Asia, and, given the nature of those conflicts, seems unlikely to diminish in the coming years. Almost 7 million Syrians were internally displaced in 2015, with another 4 million having fled to neighboring countries and others moving to Europe. In Iraq more than 3 million people were displaced internally in 2014–2015 (IOM, 2015), with a further 205,000 Iraqi refugees living outside of the country (ECHO, 2015). The violence in Yemen displaced 1.5 million within the country (IDMC, 2015) and another 100,000 as refugees (UNHCR, 2015b). Hundreds of thousands from South Sudan, Darfur, and Central African Republic were also newly displaced in 2015. According to the UN High Commissioner for Refugees (UNHCR), during 2016, "10.3 million people were newly displaced by conflict or persecution. This included 6.9 million individuals displaced within the borders of their own countries and 3.4 million new refugees and new asylum-seekers (UNHCR 2017)." In 2017, more than half a million Rohingya refugees fled into Bangladesh within a six-week period. At the same time, many refugees and internally displaced persons around the world are in protracted situations; the United Nations High Commissioner for Refugees reports that the median period of displacement is 17 years, with many Somalis, Colombians, Congolese, Afghans, and others having been displaced for even longer (UNHCR, 2015a).

Understanding not only why people are displaced but also when, where, and how they move is crucial to effective prevention and response of mass displacement. Much scholarship on migration and displacement promotes mixed-methods approaches (e.g., Graham, 1999; Silvey & Lawson, 1999; Elwood, 2010; Martin el al., 2014), which take into account the interactions between macro-level political, social, demographic, environmental, and economic factors, and related proximate causes of displacement to understand specific cases (Lindley, 2011; Lischer, 2014). One of the early efforts to relate underlying drivers of displacement with more proximate causes was Schmeidl (1997) who looked at root causes, proximate conditions, and intervening factors as potential determinants of refugee migration, with a particular emphasis on the role played by economic underdevelopment, human rights violations, ethnic and civil conflicts, external intervention, and interstate wars. In addition, she examined the impact of "flight facilitators," including migration networks and geographic proximity, and physical obstacles to movement (such as jungles or deserts). She found that underlying economic underdevelopment and population pressures have minimal impact on predicting displacement, but rather that the level and type of violence determine the likelihood and size of refugee flows. While Schmeidl (1997) examined refugee flows, Naude (2010) looked more broadly at patterns of international migration in sub-Saharan Africa and found that violent conflict and GDP growth differentials have the largest impacts on international migration in the region. Naude concluded that international migration from sub-Saharan Africa is both an adapting and mitigating strategy in the face of conflicts and economic stagnation (p. 350). Moore and

Shellman (2004) find that the magnitude of genocide and politicide significantly increase both the likelihood and magnitude of forced migration. They also suggest that past movements generate new ones either by providing a network that reduces the cost of moving for subsequent persons or by increasing the costs of staying by its disruptive effect on the society from which people flee (p. 132).

Melander and Oberg (2007) looked at magnitude and scope of conflict as determinants of forced migration and found that the intensity of armed conflict is not significantly related to the number of forced migrants. Rather they suggest that "the threat perceived by potential forced migrants is more related to where the fighting is taking place, than to the overall intensity of the fighting" (p. 157). Salehyan and Gleditsch (2006) find that the probability of violent conflict is more than three times higher in source countries than in receiving ones.

A number of studies have considered the role of intervening factors in explaining refugee flows. Massey (1988) introduces the idea of migration networks in the destination countries as contributing to forced migration, while Clark (1989) identified five intervening factors that may affect refugee outflows, including the existence of alternatives to international flight within the country, obstacles to international flight, expected reception in the asylum countries, patterns of decision-making among potential refugee groups, and seasonal factors.

Schmeidl and Jenkins (1996) discuss some of the problems of timing, where long-term or root causes may occur years before the exodus, while medium-term (or proximate) causes may occur only months beforehand. They argue: "[T]riggering events are the most difficult to place. Theoretically, they would occur almost simultaneously with, or only days before, flight but most conventional methods are unable to evaluate the close timing of triggering events" (Schmeidl & Jenkins, 1996, p. 6). They also underscore the importance of triggering events: "for policy purposes, triggering events are critical in preparing for emergency relief" (p. 6).

Households generally influence decisions as to when, where and how members will move. As Melander and Oberg (2006) explain: "The dominant explanatory model in research on both voluntary and forced migration is a decision theoretic argument, where the actors make their choices based on an assessment of the threat they experience" (p. 130) . In looking at determinants of voluntary migration, it is the expected net economic returns from migration that drive the decision about whether or not to relocate (Harris & Todaro, 1970). Research has also found that younger people are more likely to migrate for economic reasons since they have more to gain in view of their longer expected lifespan. Moreover, the previous migration of friends and relatives creates a network that facilitates subsequent migration (Hatton & Williamson, 2002).

In this vein, Davenport et al. (2003) posit that forced migrants make their decisions to abandon their homes in terms of their individual assessments of threats to personal integrity on the basis of observable information in their environment. Melander and Oberg (2006) look beyond the question of why people move to analyze the impact of forced migration flows on those that remain behind. Rather than finding that the departure of forced migrants leads to increased future flows, they find that the magnitude of flows declines over time (p. 130). They note:

> Our impression is that the more a village or town has suffered from civil war and deprivations, and the more people as a consequence have left already, the more resilient and determined to stay on is the remaining population that one encounters amongst the bullet-ridden houses and overgrown gardens. Alternatively, those that one meets in such places appear to be the oldest and weakest, with the most limited possibilities to relocate. (p. 134)

Thus, leading indicators of forced migration range from macro-level political, security, economic, social, religious, cultural, and environmental indicators to micro-level material measures that determine whether individual households have the resources and motivation to leave their homes and meso-level factors that interfere or facilitate movement (Government Office for Science, 2011). While these frameworks are useful in explaining the reasons that people stay or go in the context of conflict and other crises, they do not adequately capture the diversity in movements that occur in these situations. More effective early warning of displacement must provide greater perspective on when people move, where they go, with whom they move, what modes of transport they take, and other similar factors that determine mobility patterns in situations of conflict and repression. At this stage, no system of this type exists. One major obstacle has been access to relevant, timely local and regional data that can be incorporated into a flexible model of forced migration.

Capturing Movement Likelihood Using an Opportunity Index

This section focuses on a project begun in 2013, with funding from the National Science Foundation, MacArthur Foundation, Georgetown's Massive Data Institute and Office of the Senior Vice President for Research, and the Canadian Social Science and Humanities Research Council, on the use of big data to help identify patterns of mass displacement and begin gathering relevant data for eventual forecasting of mass displacement in the context of complex humanitarian emergencies. An exploratory study, the project enabled us to establish a network of social and computational scientists, policy makers and practitioners to explore the feasibility of developing a forecasting system using various forms of print and social media that contain indirect indicators of movement and give hints as to when and where people may relocate.

Many of the existing models of displacement, as referenced above, are linear ones, but causality is seldom linear. Moreover, the analysis of forced migration invariably begins with conditions in communities and countries of origin (the "push factors") but seldom adequately take into account the "pull" factors in likely places of destination, nor the impacts of transit points. Our understanding of the dynamics of forced migration takes into account the ways in which each factor in the model at each relevant geographic location affects and is affected by displacement and each other, creating new dynamics as the drivers of movement interact with one another.

To address this problem, we reconceived the linear models as causal loop diagrams to capture the interactions among the various factors. This approach is in keeping with the systems dynamic modeling that supports feedback loops and endogenous relationships among complex variables. Macro-level factors, such as violence and environmental degradation, interact with meso-level factors, such as transport, and communications and social networks in destination communities, and micro-level ones, such as household demographic characteristics. The causal loop demonstrates the interactions; violence, for example, may increase or decrease as relief becomes available, but relief may also increase or decrease as violence shifts in intensity and scope. In addition, the very availability of relief in particular locations may determine where people move. Furthermore, more resilient households may have greater capacity to cope with these developments (or to move) than more vulnerable ones (which may be trapped in place).

While understanding these macro-, meso-, and micro-level drivers of displacement is essential, they still do not help us fully understand why some people choose to move at certain times while others choose to move at different times or not move at all. To gain a better understanding of what triggers mass displacement in certain situations and not in others, we combined our causal analysis with dread threat theory. The pioneering work of Slovic, Fischhoff, and Lichtenstein (2000) underlies much of modern risk perception research. They employed questionnaires to characterize and quantify how ordinary people assess the risk of many hazardous activities and technologies ranging from riding bicycles to living near nuclear power plants. This research evolved into what is commonly referred to as dread threat theory (Starr, 1969; Slovic, 1987; Slovic, 2000; Slovic, Kunreuther, & White, 2000; Slovic, Fischhoff, & Lichtenstein, 2000). This research identified a heterogeneous list of "fright factors," to measure people's responses to safety questions. Because we are often dealing with situations of persistent threat, we add a dynamic element to dread threat theory—the menacing context that emerges when a dread threat persists and requires a community to reorganize its life to mitigate consequences of threat—and employ data from journalistic, ethnographic, biographical, and autobiographical sources. The concept of menacing context has evident value in analyzing the determinants of forced migration because it links situational factors to decision-making as well as macro, meso, and micro levels of analysis through local perceptions of, and responses to, dread threat (Collmann et al., 2016). Also, recognizing that there are strong pull factors even in cases of forced migration—including better economic and educational opportunities, greater physical safety, better access to shelter, food, and other commodities, etc., in other locations—our analysis led us to identify a need to construct an "opportunity" index that will allow early warning systems to measure the attraction of various alternative options.

To validate these theoretical perspectives, we conducted interviews with Iraqi and Syrian refugees, internally displaced persons (IDPs), local host populations in the neighborhoods in which refugees and IDPs are located, as well as stakeholders in Jordan, Lebanon, Turkey, and Iraq. We used a snowball sample, recruiting interviewers that represent a wide range of socio-demographic and geographic (where they come from and where they are currently living) characteristics, and then asked them to recruit people for the interviews. Though not random, the resulting snowball sample resembles the total population of interest. All of the interviews cover the same issues but the questions are open-ended, allowing the respondents to move the discussion into areas that they consider most important.

These interviews are ongoing but have already been used in two respects. Initially, the results were used to help inform the development of the causal loop diagram and the dread threat model with stakeholder perspectives. The computer scientists extensively interviewed the researchers about the determinants of displacement in the two case studies. The information conveyed was based primarily on the stakeholder interviews. The interviews with the affected populations are used to validate the findings coming out of the big data analysis that will be described further in the next sections. As we refine the big data analysis, we conduct further interviews with Iraqis and Syrians to evaluate the individual and household dynamics of how, when, and using what criteria they decided to leave their communities. These new interviews allow us to study more recently displaced households, including recent movements to Europe and elsewhere.

Using Big Data to Identify Determinants and Triggers of Mass Displacement

Along with extending the theoretical framework of forced migration, this project gave us the opportunity to begin analysis of local print and social media, specifically Twitter, to identify the changing dynamics of events and perceptions that may directly or indirectly trigger displacement in and from Syria and Iraq since 2011. We used an archive of more than 700 million publicly available open-source media articles that has been actively compiled since 2006. News articles are added to the Expandable Open Source (EOS) database at the rate of approximately 100,000 per day by automated scraping of Internet (web-based) sources in 46 languages across the globe (Singh and Pemmaraju, 2017). We also compiled a database containing over 1.5 billion tweets in English and Arabic from organizations and individuals that regularly post on developments in Iraq and Syria, and on relevant hashtags, including ones related to ISIL. Using newspaper and social media data begins to give us a glimpse into what people are talking about, what their perception of different events and conditions are, and whether or not concern is increasing or decreasing about various factors related to forced migration.

To determine which ideas, types of events, and topics are correlates of movement or correlates to direct indicators that may not be available during different crisis situations, we have begun using statistics compiled by the United Nations High Commissioner for Refugees, the International Organization for Migration, the Office for the Coordination of Humanitarian Affairs, and the Internal Displacement Monitoring Centre. Demographic data and economic indicators can also be drawn from standard sources, such as the UN Human Development Index and the World Bank. These data can serve as indirect indicators/variables in the context of migration. Therefore, we need to understand the relationship between these known variables and the variables extracted from noisy, partial, open-source big data. Are they well correlated or is there a limited relationship between them? We can also try to correlate big data variables to interview data collected in different volatile regions around the world. Because interviewing is not scalable, if we can find strong correlates in big data variables, we can use them as proxies for traditional interview variables that may be difficult to obtain in certain unstable regions of the world.

So the primary question becomes—how do we identify meaningful forced migration-related variables from big data sources? In the previous section, we highlighted a number of factors that influence an individual's decision to migrate or not during conflict. Our approach hinges on understanding: (1) the changing dynamics of each factor in a particular location; and (2) the importance of each factor within a particular location or community. We can measure both of these by analyzing the changing newspaper and social media content related to these factors. To accomplish this, we begin the process by identifying relevant documents using state of the art information retrieval techniques, extracting useful structured data representations, i.e., sketches of the data, and then using these data representations to construct variables for use in a dynamic forced migration model (Wei et al., 2014; Wei and Singh, 2017). For example, creating a vector of words about violence and computing the frequency of these words across newspaper articles each day can be used as the basis for a time series variable that captures the changing dynamics of violence in a particular location. These changing dynamics may be a strong indirect indicator of movement in certain conflict areas. Preliminary work that correlates migration words in text to levels of migration based on bi-weekly Iraqi population movement data from the International Organization of Migration shows promise. Another type of data sketch may be a semantic graph that contains words and phrases as nodes and relationships based on co-occurrence of these words and phrases in articles or tweets. This type of graph can be useful for identifying frequently occurring groups or clusters of words/discussions of local and regional topics of interest. While it is also possible to generate administrative variables from these big data sources, we believe that researchers need to explore big data in new ways and generate new types of variables to gain insight that differs from values that can be determined in other ways. Here we describe interesting variables that we hypothesize will help our understanding of movement.

Events: An *event* is something that happens at a particular time and location, e.g., a bombing in Anbar on January 10, 2015. A targeted event is an event in a particular location that is associated with a particular domain or topic of interest to the user, e.g., politics, violence, football, etc. An interesting variable extracted from big data is the frequency of targeted events. Doing so allows us to compute a time series containing the number of targeted events related to topics correlated to forced migration each day. The frequency of different types of discovered events and the topics associated with these events can themselves also be used as indirect indicators of forced migration. Because of this, we are also interested in mapping the detected events and their topics to different factors associated with forced migration. This approach allows us to integrate knowledge from interviews with knowledge from social media to gain a more accurate picture of the situation.

Perception: In order to understand whether or not people will choose to migrate, it is important to understand their perceptions about relevant direct and indirect indicators, e.g., wages, schools, etc. Perceptions can be measured in different ways. Three that are important in the context of migration are tone (sentiment), stance (position), and emotion. An important research direction is to learn to identify tone, stance, and emotion from social media and newspaper content so that perception can be more accurately captured. Currently, projects use the notion of sentiment as a proxy for perception. Sentiment indicates the opinion of the author of the article or tweet. Sentiment can be positive, negative, or neutral. Our preliminary work suggests that sentiment related to ISIS changed over a one-year period when different events occurred. What is also evident is that the sentiment is not always the same in different languages and/or locations (Singh et al., in preparation). Stance involves determining the position of a tweet or article toward a particular entity. For example, is a tweet pro-ISIL or anti-ISIL? Finally, emotion considers whether the tweet or article contains emotional content, e.g., happy, sad, angry, etc. An important future direction is to use perception determined from open source data to further investigate dread threat variables on a broader scale. For example, if other sources of information suggest an increase in dread threat levels in Iraq over time, we can determine if that same increase occurs on Twitter. If we are able to map variable values obtained from other sources to variables extracted from tweets, we may be able to further our understanding of the drivers and triggers of forced migration and see the escalation of dread threat levels before large-scale displacement occurs.

Buzz: Buzz represents the amount of interest in a topic. Buzz may be popular and trending, but it may have low values. What is interesting is the variation of buzz strength of a topic over time. This buzz strength will be based on the frequency of occurrence of phrases in articles and tweets for a particular location. One can imagine using a heat map to see the buzz of different topics in different locations. This can give us immediate insight into the distribution of the factors of interest in a particular region. This distribution is vital for understanding the specific factors that may be more important in different parts of the world.

To date, we have focused on *violence* related events, buzz, and perception variables since those appear readily in newspaper and social media data. We used semi-supervised machine learning methods to identify a set of terms and phrases that represent the *violence* topic. We then extracted daily buzz, event, and perception variables for specific locations in Iraq during 2013–2014 and demonstrated how they correlated to important events associated with forced migration (event and perception index) (Wei et al., 2016), and death counts in Iraq (buzz index) (King, 2016). The same process can be used to capture other indirect indicators of forced migration. Our project will monitor relevant indicators using EOS and Twitter, and incorporate them into an agent-based simulation tool that models possible movements using both theoretical and data-driven knowledge about individual and contextual conditions and changes to conditions. For this to be effective over time, we need to identify the variables that serve as reasonable proxies for pull and push factors that cannot be easily determined in other ways.

Tools and Analytics

As mentioned in the introduction, we must have tools to help policy makers understand the impact of not acting in certain situations. So far we have focused our discussion on an early warning tool that issues alerts when shifts in movement patterns or new movements seem to be occurring. Another important tool is a simulation tool that can help policy makers analyze the evidence (newspaper articles, social media posts) that triggered the warning and simulate different "what if" scenarios, enabling policy makers to see possible outcomes of different interventions.

Early Warning Tool: An early warning tool should be capable of using indicators drawn from different data sources (many real-time sources) within a dynamic theoretical model to alert decision-makers to likely changes in patterns of displacement. In some cases, the displacement will be new, but in many situations, the alert will mark potential shifts in movements. The alert system should go well beyond the binary decision to move or stay. It should seek to provide information to decision-makers on who will move (i.e., what are their demographic and socioeconomic characteristics), in what numbers, from where, to where. It should also present policy makers with the evidence used to generate the alert and a way for the policy maker to input the strength, reliability, and timeliness of the evidence, thereby allowing the tool to learn from human analysis of the evidence.

Simulation Tool: Simulation tools can provide decision-makers the capacity to test responses to patterns of movements under varying scenarios. For example, if displacement is related to increasingly more severe food insecurity, decision-makers could test various scenarios involving the delivery of food to at-risk populations—including purchase of food in neighboring countries, vouchers to enable people to buy available food, shipment of food from more distant countries, food drops, food distribution in camps, etc. We see two purposes for such scenario testing. First, it helps determine the likely results of a humanitarian action, e.g., what if food relief is dropped at a particular location? Second, it gives insight into determining the likely results of a third-party action, e.g., what if the Jordanian government closes its border with Syria?

One type of analytic tool that can be particularly helpful is a computational simulation that gives practitioners an opportunity to posit a scenario via a web user interface, run the simulation, and view the simulated results though a geographic visualization. A simulation could forecast 7–20 days ahead, based on what is known and what can be inferred. We anticipate that practitioners who had access to such a system would run many such scenarios each day, to better understand the scope of what is possible.

At Georgetown, we built a prototype of how local perception of threat in the locality drives actions to mitigate that threat, including both planned migration and unplanned flight. The simulation already developed is based on system dynamics, defined as a "computer-aided approach to policy analysis and design" that "applies to dynamic problems arising in complex social, managerial, economic, or ecological systems."[4] Simulations based on systems dynamics have several advantages, including ease of development and computational tractability, but also come with limitations on modeling the inherent economic and social diversity of human populations. In effect, systems dynamics models do not necessarily capture decision-making at the household and individual level.

By contrast, agent-based simulation of forced migration allows for modeling each individual household, where a household decides whether and when to migrate, based on its unique assets, location, social connections, time-varying perception of threat, and other factors (Edwards 2008; Kniveton et al. 2011; Kuznar & Sedlmeyer, 2005; Smith 2012). Often lost in this type of analysis, however, are the systems that may facilitate or impede the household from taking certain actions. Integrating the two models could be valuable for simulating different types of interactions. To date, no full-scale alert system or simulation platform that incorporates both analytic models exists for forced migration.

Current Relevant Uses of Different Types of Big Data

A limitation on our approach to date is its reliance on certain types of data—news articles and tweets—that rely on Internet coverage, which may not be available in countries experiencing large-scale displacement. For example, Internet penetration in Syria (29%) is much higher than in Iraq

(13%), but still low compared to the US (89%) (Internet Live Stats, 2016). Therefore, insight from social media sources may be limited for certain countries and even more for areas of a country that is mired in conflict. The key to using social media for these forms of analysis, therefore, is to identify social media sources that have high penetration in the regions of interest, particularly among those who are likely to be reliable informants of events, perception, and buzz.

Within this context, there are other forms of big data that would be valuable for informing early warning and simulation tools for forced migration. Here we mention a few. Facebook has much more robust information about social networks of individuals, demographics, and conversations than Twitter. The Data Science team at Facebook has already used Facebook data to infer lifetime migration by analyzing users' home towns and current locations (Hofleitner et al., 2013). LinkedIn data has been used to investigate labor market migration in parts of the U.S. (State et al., 2014). We can imagine using a site containing career history information to determine places where job opportunities are improving or declining.

Other important data sources for movement applications are email IP addresses (Zagheni & Weber, 2012), mobile phone data (Gonzalez et al., 2008) and satellite imagery data. Using phone GPS in conjunction with text messaging has already become an important data source for providing aid during crises, e.g., 2010 Haiti earthquake (Bengtsson et al., 2011), Rwanda migration around 2010 (Blumenstock, 2012) and for monitoring seasonal population changes in France and Portugal (Deville et al., 2014). It has also been shown to be a good source of data for monitoring the diffusion of epidemics and the effectiveness of different public health measures (Frias-Martinez et al., 2012). Mobile phones are a cheap form of communication with higher penetration than other technologies. Satellite imagery, while difficult to obtain, is a way to observe physical movement of people in crisis situations. It gives better insight into directions traveled and routes taken, making it an important future data source for determining where people are moving and where camps and aid should be placed. Concerns with these data types include their lack of availability for researchers, inconsistent amounts of data available for different regions of the world, and the risk associated with breaching personal privacy of individuals if used.

These big data sources were not originally developed for forced migration or crises applications. However, most have already been shown to be useful for disaster management applications, understanding mobility patterns, and more generally, for learning about human behavior in different situations.

Challenges and Limitations in Using Big Data

For all the benefits of big data, a number of challenges exist. First, most of these data are noisy and partial. The signal-to-noise ratio for most topics is very low. Second, the reliability of different sources, and even authors of articles/social media posts, is not clear. These data may also have significant biases. Systematic bias is very different from random error and may be hard to identify, much less compensate for. In order to effectively use big data, we must develop methods and tools to quantify and adjust for the variability in reliability and the potential high levels of bias. Third, big data population coverage varies considerably in terms of demographic and movement data. As technology continues to get cheaper and more pervasive, the utility of big data will continue to grow. Next, there is a lack of reliable ground truth data for algorithm output comparisons. While there is some knowledge about where and when people move, it is inconsistent, noisy, and not timely. In order to calibrate algorithms and understand their strengths and weaknesses, having ground truth data is important. Finally, it is difficult to integrate large numbers of sources of data that have varying temporal and spatial resolutions. Using time and GPS coordinates is the most straightforward way to combine these data, but using semantic similarity is an important future direction. A large public and/or private initiative that promotes standardization and interoperability across different distributed platforms and entities is an important direction for making traction on these large-scale challenges.

We need as granular and dynamic data as possible in order to identify relevant indicators of forced migration. The scale of migration can significantly redistribute population, within and across borders, in very short periods. Consider the crisis-driven migration of one million Syrians 2013–2015 to Lebanon, a country of 4.5 million, or over one million Syrians and others to Western Europe in 2015. As "big" as our data sources are now, they do not include information in all of the language groups needed to forecast displacement, nor are the sources sufficiently local (meaning to the community, and, even, household level) to allow us to get at the meso- and micro-level factors influencing movement, particularly in areas where social media penetration is low. Data availability will be vital for making significant progress in this area. We also need to use these data with care, considering anonymization strategies to ensure privacy and developing guidelines for the ethical uses of these personal data. While these data can be used for social good, their availability also allows for disruptive forces to use these data (Singh, 2016). We are particularly concerned that such information could be used to target people, as was done with census data during the Rwandan genocide, or to deter flight even if it is the only way for people to achieve safety. Efforts need to be undertaken to ensure that does not happen.

Conclusion

Making progress on understanding the drivers of forced migration, and developing tools to forecast when, where, how, and who will be displaced, will have a potentially profound impact on understanding and coping with future movements. Early warning holds the potential to save lives and to make humanitarian responses more effective. It would improve planning as well as directly aid potential refugees before, during, and after their exodus. Such planning can lead to action to try to avert mass displacement, help divert forced migrants from risky modes of movement (e.g., via unseaworthy boats or across landmine infested borders), and enable governments and international organizations to pre-position shelter, food, medicines, and other supplies in areas that are likely to receive large numbers of refugees and displaced persons. Although governments will not always act benevolently in the face of early warning of displacement—such warnings can also give governments more time to stop refugees from crossing onto their territory—the alternative is often chaos, with the displaced and the communities they enter left without adequate assistance or protection. Big data, if integrated responsibly and combined with available administrative data, can be the catalyst for a timely, reliable early warning system and a mobility simulation platform that identifies likely movement patterns given different policy options. Over the next five to ten years, it is important for humanitarian agencies, researchers, and corporations to work together to integrate these data and harness them to improve the situation of those forced to migrate.

Acknowledgments

We are fortunate to have a large team of contributors. We would like to acknowledge the work of Jeff Collmann, especially for his perspectives on dread threat, as well as Lara Kinne, Nili Yossinger, Abbie Taylor, Susan McGrath, and Yifang Wei. This work was supported in part by the National Science Foundation (NSF) Grant SMA–1338507, the Georgetown University Mass Data Institute (MDI), the John D. and Catherine T. MacArthur Foundation, and the Canadian Social Science and Humanities Research Council (SSHRC). Any opinions, findings, conclusions, and recommendations expressed in this work are those of the authors and do not necessarily reflect the views of NSF, MDI, the MacArthur Foundation, or SSHRC.

Notes

1. See http://www.fews.net/ and http://wcatwc.arh.noaa.gov/, respectively.

2. http://www.crisisgroup.org/en/publication-type/alerts.aspx.

3. http://www.acleddata.com.

4. As defined by the Systems Dynamics Society, http://www.systemdynamics.org/what-is-s.

References

Bengtsson, L., Lu, X., Thorson, A., Garfield, R., & von Schreeb, J. (2011). Improved Response to Disasters and Outbreaks by Tracking Population Movements with Mobile Phone Network Data: A Post-Earthquake Geospatial Study in Haiti. *PLoS Med., 8*(8): e1001083-10.1371/journal.pmed.1001083.

Blumenstock, J. E. (2012). Inferring Patterns of Internal Migration from Mobile Phone Call Records: Evidence from Rwanda. *Information Technology for Development, 18*(2), 107–125.

Clark, L. (1989). *Early Warning of Refugee Flows*. Washington, DC: Refugee Policy Group.

Collmann, J., Blake, J., Kinne, L., Dillon, R., Bridgeland, D., Martin, S., et al. (2016). Measuring the Potential for Mass Displacement in Menacing Contexts. *Journal of Refugee Studies, 29*, 273–294.

Davenport, C. A., Moore, W. H., & Poe, S. C. (2003). Sometimes You Just Have to Leave: Domestic Threats and Forced Migration, 1964–1989. *International Interactions, 29*(1), 27–55.

Deville, P., Linard, C., Martin, S., Marius, G., Stevens, F., Gaughan, A., et al. (2014). Dynamic Population Mapping Using Mobile Phone Data. *Proceedings of the National Academy of Sciences of the United States of America, 111*(45), 15888–15893. doi:10.1073/pnas.1408439111.

Edwards, S. (2008). Computational Tools in Predicting and Assessing Forced Migration. *Journal of Refugee Studies, 21*(3), 347–359.

Elwood, S. (2010). Thinking Outside the Box: Engaging Critical Geographic Information Systems Theory, Practice and Politics in human Geography. *Geography Compass, 4*(1), 45–60.

European Commission Humanitarian Aid and Civil Protection (ECHO). (2015, September). Fact Sheet—Iraq. http://ec.europa.eu/echo/files/aid/countries/factsheets/iraq_en.pdf?utm_source=viva&utm_medium=print&utm_campaign=viva-July-August-2015.

FEWS.net. Famine Early Warning System. Available at http://www.fews.net.

Frias-Martinez, V., Rubio, A., and Frias-Martinez, E. (2012). Measuring the Impact of Epidemic Alerts on Human Mobility. Pervasive Urban Applications—PURBA, Newcastle, UK.

Frias-Martinez, E., Williamson, G., & Frias-Martinez, V. (2011). *An Agent-Based Model of Epidemic Spread Using Human Mobility and Social Network Information.* International Conference on Privacy, Security, Risk and Trust (PASSAT), Boston, MA, pp. 57–64.

Government Office for Science (UK). (2011). Foresight: Migration and Global Environmental Change. Final Project Report. London.

Graham, E. (1999). Breaking Out: The Opportunities and Challenges of Multi-Method Research in Population Geography. *Professional Geographer, 51*(1), 76–89.

Harris, J., & Todaro, M. (1970). Migration, Unemployment, and Development; A Two-Sector Analysis. *American Economic Review, 60*, 126–142.

Hatton, T. J., & Williamson, J. G. (2002). Out of Africa? Using the Past to Project African Emigration Pressure in the Future. *Review of International Economics, 10*(3), 556–573.

Gonzalez, M. C., Hidalgo, C. A., & Barabasi, A.-L. (2008). Understanding Individual Human Mobility Patterns. *Nature, 453*(7196), 779–782.

Government Office for Science, United Kingdom. (2011). Foresight: Migration and Global Environmental Change: Future Challenges and Opportunities. Final Project Report. London. https://www.gov.uk/government/uploads/system/uploads/attachment_data/file/287717/11-1116-migration-and-global-environmental-change.pdf.

Hofleitner, A., Chiraphadhanakul, T., & State, B. (2013, December 17). Coordinated Migration. Facebook Data Science Team. https://www.facebook.com/notes/facebook-data-science/coordinated-migration/10151930946453859.

Internal Displacement Monitoring Centre (IDMC). (2015, August). Yemen IDP Figures Analysis. http://www.internal-displacement.org/middle-east-and-north-africa/yemen/figures-analysis.

IDMC. (2016). Global Report on Internal Displacement. http://www.internal-displacement.org/globalreport2016.

International Organization for Migration (IOM). (2015, November 20). Displacement Continues in Iraq Amid Return Movement. http://iomiraq.net/article/0/displacement-continues-iraq-amid-return-movements-iom.

InternetLiveStats. (2016, July 8). Internet Users By Country. http://www.internetlivestats.com/internet-users-by-country.

King, J. (2016). *Methods to Overcome Challenges When Learning Arabic Word Embeddings for Text Mining Tasks* (Undergraduate senior thesis in Computer Science). Georgetown University, Washington, DC.

Kniveton, D., Smith, C., & Wood, S. (2011). Agent-based Model Simulations of Future Changes in Migration Flows for Burkina Faso. *Global Environmental Change, 21*(Supplement 1), S34–S40.

Kuznar, Lawrence A., & Sedlmeyer, R. (2005). Collective Violence in Darfur: An Agent-Based Model. *Mathematical Anthropology and Cultural Theory: An International Journal, 1*(4), Paper 1105.

Lindley, A. (2011). Between a Protracted and a Crisis Situation: Policy Responses to Somali Refugees in Kenya. *Refugee Survey Quarterly, 30*(4). https://doi.org/10.1093/rsq/hdr013.

Lischer, S. K. (2014). Conflict and Crisis Induced Displacement. In E. Fiddian-Quasmiyah, G. Loescher, K. Long, & N. Sigona (Eds.), *The Oxford Handbook of Refugee and Forced Migration Studies.*

Martin, S., Weerasinghe, S., & Taylor, A. (2014). *Migration and Humanitarian Crises: Causes, Consequences and Responses*. New York, NY: Routledge.

Massey, D. (1988). Economic Development and International Migration in Comparative Perspective. *Population and Development Review, 14*, 383–413.

Melander, E., & Oberg, M. (2006). Time to Go? Duration Dependence in Forced Migration. *International Interactions, 32*, 129–152.

Melander, E. and Oberg, M. (2007). The Threat of Violence and Forced Migration: Geographical Scope Trumps Intensity of Fighting. *Civil Wars, 9*(2), 156–173.

Moore, W. H., & Shellman, S. M. (2004). Fear of Persecution: A Global Study of Forced Migration, 1952–1995. *Journal of Conflict Resolution, 40*(5), 723–745.

Naude, W. (2010). The Determinants of Migration from Sub-Saharan African Countries. *Journal of African Economies, 19*(3), 330–356.

NOAA. National Oceanic and Atmospheric Administration National Tsunami Warning Center. http://ntwc.arh.noaa.gov.

Office for the Coordination of Humanitarian Affairs (OCHA). (2015, November). Syria Crisis: Country-Based Pooled Funds. http://www.unocha.org/syria.

Salehyan, I., & Gleditsch, K. S. (2006). Refugees and the Spread of Civil War. *International Organization, 60*(2), 335–366.

Schmeidl, S., & Jenkins, J. C. (1996). Issues in Quantitative Modelling in the Early Warning of Refugee Migration. *Refuge: Canada's Periodical on Refugees, 15*(4), 4–7.

Schmeidl, S. (1997). Exploring the Causes of Forced Migration: A Pooled Time-Series Analysis, 1971–1990. *Social Science Quarterly, 78*(2, June), 284–308.

Silvey, R., & Lawson, V. (1999). Placing the Migrant. *Annals of the Association of American Geographers, 89*(1), 121–132.

Singh, L. (2016). Data Ethics—Attaining Personal Privacy on the Web. In J. Collmann & S. A. Matei (Eds.), *Ethical Reasoning in Big Data: An Exploratory Analysis* (pp. 81–90). New York: Springer.

Singh, L. and Pemmaraju, R. (2017). *EOS: A Multilingual Text Archive of International Newspaper & Blog Articles*. IEEE International Conference on Big Data (BIGDATA). Boston, MA.

Singh, L., Wei, Y., Kirov C., Taylor, A., Yossinger, N. & Martin, S. (In preparation). Tracking sentiment of ISIS on Twitter.

Slovic, P. (1987). Perception of Risk. *Science, 236*, 280–285.

Slovic, P. (2000). *The Perception of Risk*. Sterling, VA: Earthscan Publications.

Slovic, P., Fischhoff, B., & Lichtenstein, S. (2000). Facts and Fears: Understanding Perceived Risk. In P. Slovic (Ed.), *The Perception of Risk* (pp. 137–153). Sterling, VA: Earthscan Publications.

Slovic, P., Kunreuther, H., & White, G. (2000). Decision Processes, Rationality and Adjustment. In P. Slovic (Ed.), *The Perception of Risk* (pp. 1–31). Sterling, VA: Earthscan Publications.

Smith, C. D. (2012) Assessing the impact of climate change upon migration in Burkina Faso: an agent-based modeling approach. (Doctoral dissertation). University of Sussex, UK. Available online via Sussex Research Online, http://sro.sussex.ac.uk.

Starr, C. (1969). Social Benefit Versus Technological Risk. *Science, 165*, 1232–1238.

State, B., Rodriguez, M., Helbing, D., & Zagheni, E. (2014). Highly skilled immigrants are losing interest in the United States: LinkedIn data. https://www.linkedin.com/pulse/highly-skilled-immigrants-losing-interest-united-states-maha-hamdan.

United Nations High Commissioner for Refugees (UNHCR). (2016). *Global Trends in 2015*. Geneva: UNHCR.

United Nations High Commissioner for Refugees (UNHCR). (2015a, June). World at War: Global Trends—Forced Displacement in 2015.Geneva: UNHCR. http://www.unhcr.org/556725e69.html.

United Nations High Commissioner for Refugees (UNHCR). (2015b, October 5). Yemen Situation: Regional Refugee and Migrant Response Plan, October–December 2015. http://reliefweb.int/report/yemen/yemen-situation-regional-refugee-and-migrant-response-plan-october-december-2015-0.

United Nations High Commissioner for Refugees (UNHCR). (2017). Global Trends: Forced Displacement in 2016. Geneva: UNHCR. http://www.unhcr.org/globaltrends2016.

Wei, Y. and Singh, L. (2017). Location-Based Event Detection Using Geotagged Semantic Graphs. Workshop on Mining and Learning with Graphs (MGL) at the ACM International Conference on Knowledge Discovery and Data Mining (KDD). Nova Scotia, Canada.

Wei, Y., Singh, L., Gallagher, B., & Butler, D. (2016). *Overlapping Target Event and Storyline Detection of Online Newspaper Articles*. IEEE International Conference on Data Science and Advanced Analytics. Montreal, Canada.

Wei, Y., Taylor, A., Yossinger, N. S., Swingewood, E., Cronbaugh, C., Quinn, D. R., et al. (2014). *Using Large-scale Open Source Data to Identify Potential Forced Migration*. Paper presented at and published for the 2014 KDD Workshop on Data Science for Social Good. http://dssg.uchicago.edu/kddworkshop.

Zagheni, E., & Weber, I. (2012). You Are Where You Email: Using E-mail Data to Estimate International Migration Rates. *Proceedings of the ACM Web Science Conference 2012*, Oxford, UK, 348–358.

III Information Policies and the Displacement Research Agenda

10 Information Policies and Displacement

Carleen F. Maitland

Information policies play a crucial role in meeting the extensive and changing information needs of displaced persons, humanitarian organizations, and forced migration researchers alike. They not only define how data and information are collected, shared, and used, but may also define which technologies are used in these processes. In turn, the overall costs and benefits of technological change, as well as to whom they accrue, may be determined. Given these direct and secondary effects, information policies provide answers to fundamental questions, such as: Whether and how refugees can access their own data? Which technologies should and can be used in supporting refugees?

It is not always clear which organizations have the responsibility or right, or even the capacity, to develop and enforce these policies. Jurisdictions vary, in some cases resulting in policies having effects across multiple levels (international, national, organizational, and individual). In other cases, policies remain unspecified, as the correct course of action remains unclear or the prospect of taking responsibility, not only for formulation but also enforcement, is daunting.

Where information policies do emerge, they both shape and are shaped by technological change. They define answers to questions related to serving the displaced, such as: In a particular location, who can use wireless spectrum, biometrics, and collect data from children? To what end? At the same time, new technologies drive the need for new policies to control their effects. Questions include: What policies are needed to ensure efficient use of spectrum, and balance the public's interest in security with the privacy needs of individuals? For displaced persons, the impacts of technological change are both amplified and mitigated by information policies.

Through an inductive analysis of the policy issues raised throughout this volume, three priority policy domains are identified: networks, refugee rights, and data management. This chapter examines how information policies in these three domains emerge at the international, national, organizational, and even individual levels. The analysis is conducted within a socio-technical frame, investigating the centrality of information policies in defining limits to, as well as the potential benefits of, technology use. It also provides critical insight into the ways technological change is driving the demand for information policies.

In what follows, a background section presents the results of the inductive analysis, further articulating the jurisdictional levels and then arguing for the delineation of the network, refugee rights, and data management domains. Following this, the policy issues for each jurisdiction and domain are presented. These discussions include a synthesis of policy issues raised by the authors, but also identify related issues and offer greater insight into policies and policymaking bodies. Finally, the chapter concludes with a research agenda for information policies related to ICTs for refugees and the displaced.

Background

New information policies emerge when technological change either makes existing policies inefficient or irrelevant, or creates a need for constraints or guidance unlikely to be fulfilled through other means. The new policies must balance the need to be forward looking while maintaining relevance to the current state-of-the-art. An important criterion for assessing this balance is a policy's robustness to further technological change (Bauer, 2014; Bauer & Bohlin, 2007). Failing to create this balance results in policies quickly becoming out of date, limiting innovation, or becoming irrelevant, unenforceable, or ignored.

As opposed to a narrow focus on *public policy*, wherein policymaking is exclusively a public-facing activity, affecting many citizens, here policymaking is treated as an activity occurring not only in public political bodies, such as the UN General Assembly and national legislatures, but also in organizations and by individuals. Together, these various entities engage in the policymaking that shapes the experiences of refugees and the displaced.

Policy Jurisdictions

Policy jurisdictions include both where and by whom policies are made and/or where they have impact. For example, a policy made by the UN General Assembly may have its greatest impact at the national level, where it serves as the basis for national legislation. Table 10.1 identifies the *primary* jurisdiction of impact, derived through a careful review of this volume's chapters. It also points to the various entities responsible for or having the necessary power to develop and enforce policies.

The review identified many information policymaking jurisdictions. In part I, "Legal, Social, and Information Science Perspectives," the discussions presented by Ruffer (chapter 2) and Kingston (chapter 3) primarily focus on international legal instruments, the effects of which are strongest at the national level. In contrast, Maitland (chapter 4) and Fisher (chapter 5) demonstrate the implications of policies at the organizational and individual levels. In part II, "Technical Perspectives," the four chapters also point to international policies, but have their primary impacts at the national and organizational levels.

The positions of various policymaking entities, typically organizations, within the international, national, and organizational jurisdictions are depicted in figure 10.1. Of interest here are the ways the various levels of law and policymaking interact. For example, laws generated at the international level are then transposed or translated into national laws by (most) member states. Multilateral bodies, such as the European Union, may also transpose laws, adding both a collective voice and an additional layer of "translation" into the law-making process.

At the same time, UN specialized agencies serving the displaced, such as UNHCR, WFP, and WHO, as well as technical agencies such as the International Telecommunications Union, must also comply with UN legal frameworks. Depending on their governance structures, these agencies may also make international policies for their more narrow operational domains. In some cases, a narrow mandate may result in less broadly impactful policies than compared with those generated by agencies dealing with cross-cutting issues such as human rights. For this reason, the specialized agencies are shown as straddling all three levels and are discussed in any of these jurisdictions.

At the organizational level, various entities, such as governmental organizations, including Ministries of Health and Education, create policies

Table 10.1
Information policy issues and their jurisdictions

Chapter	Policy issues	Primary jurisdiction
2. Informational components of refugee status determination.	–National policies stipulating valid forms of information	–National
	–National policies stipulating who can present information	–National
	–National policies concerning collection of primary documents in RSD	–National
	–Organizational policies determining practices in document collection in RSD	–Organizational
3. Statelessness, identity and biometrics.	–National policies for issuing identity documents for citizens	–National
	–National policies for issuing identity documents to refugees	–National
	–National policies directing agencies to accept refugee identification	–National
	–Organizational policies for the management of biometric data	–Organizational
4. Information sharing and multi-level governance in refugee service provision.	–Policies related to information technologies and data sharing as critical program elements (M&E)	–Organizational
	–Organizational prohibitions on/requirements for sharing data	–Organizational
5. Information sharing by and among refugees.	–Individual policies on data management	–Individual
	–Policies protecting data privacy of mobile phones	–National
6. Cellular and wireless infrastructures.	–Telecom policies that grant exclusive use of spectrum	–National
	–Policies that inhibit infrastructure provision by the UN	–Organizational
7. Information systems and technologies.	–Policies on data protection	–Organizational
	–Information security policies	–Organizational
	–Policies for sharing data	–Organizational
8. GIS and displacement.	–Policies related to sharing of GIS data collected by governments	–National
	–Policies related to sharing of spatial data collected by organizations	–Organizational
	–Policies related to promotion of mapping	–Organizational
9. Data analytics and displacement.	–Open data policies that enable access	–National
	–Policies that promote use of externally generated predictions into organizational decision-making	–Organizational

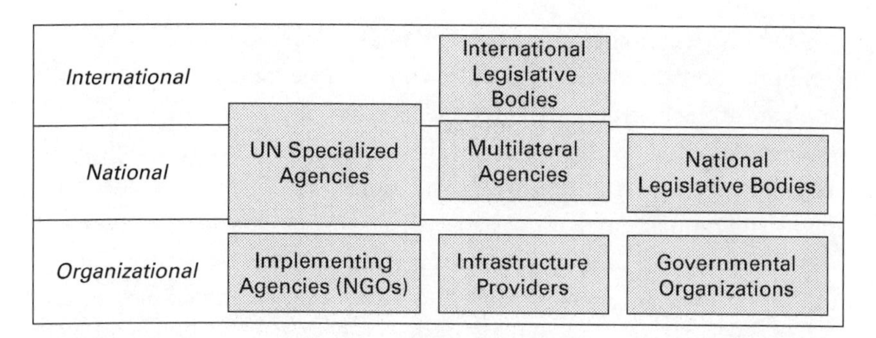

Figure 10.1
Jurisdictional levels for refugee-related information policy making.

informed by national legislation, but their jurisdiction typically is restricted to their domain of operation. This is not to imply a solely top-down flow of influence, as UN agencies and indeed international law are also shaped by national and organizational policies and legal frameworks (see e.g., Scheel & Ratfisch, 2014).

Finally, while policymaking is often investigated at organizational or national levels, today individual displaced persons will formulate, adhere to, and enforce their own individual information policies. In privacy scholarship, such actions have been referred to as "privacy self-management" (Solove, 2013). However, similar to organizations, these individual information policies go beyond privacy, to include storage, curation, and security as well.

Policy Domains

As reflected in this volume, technological change is shaping network technologies and architectures, refugee rights, and data management. Network technologies serve as the basis not only for communication, but also for data flows and information systems affecting the lives of the displaced and the operations of humanitarian organizations. Diverse devices, ranging from the cameras on mobile phones that have enabled the creation of portable family photo albums to the iris scanners used in registration, capture data to be stored, managed, and potentially shared. Policies at various levels shape the availability of networks and the types of devices connecting to them.

Similarly, technological change has implications for information policies in the domain of refugee rights. While refugee rights policies typically originate at the international level, the intersection of these rights and new and emerging ICTs will occur at all levels. Refugee rights articulated at the national level will define basic uses and limits to ICTs for national governments and humanitarian organizations.

The third policy domain considers issues of data management across all levels and stages in the refugee life cycle. The widespread use of data capture technologies has created a greater focus on data management. Policies specify how, when, and why data can be captured, stored, and shared. Here, the term "data management" is used, in contrast to the more popular notions of data protection or data privacy. This choice is meant to emphasize the broader range of activities needed to protect not only data, but also the people they represent.

Jurisdiction 1: International

Even though the primary jurisdictions for the information polices of interest here are at the national and organizational levels, international legal frameworks are important for two reasons. Often established by the UN, they are significant both for instituting what can be taken as a global legal norm and for their coordination function. The coordination function incentivizes nation-states to act in roughly the same timeframe, allowing the emergence of best practices for translating the frameworks into national law (Abbott, Snidal, & Abbott, 1998). For the information environment of refugees, these legal frameworks have both indirect effects, in shaping the embedded technical base, and direct effects, through establishing rights. While these rights may or may not be recognized, they do provide direction as to the types of information allowed to play a role in gaining certain protections (e.g., identity documentation).

Network Access

At the international level, legal frameworks with indirect effects on the technologies available to refugees and the displaced are seen in Radio Regulations, a treaty of the International Telecommunications Union. The treaty, ratified by 200 of the Union's Member States,[1] shapes coordination on, and the availability of, spectrum for a wide variety of wireless technologies.

While Schmitt et al. (this volume) call for changes in national regulations, it would be an oversight to ignore the potential role of the ITU in this regard. The ITU informally shapes the interpersonal networks through which policy diffusion, and, in turn, a limited form of harmonization, sometimes occur.[2]

Refugee Rights

International laws affecting refugees' information environments range from the very specific, such as those defining the term "refugee," to the general, such as those concerned with human rights. The former include the UN Convention Relating to the Status of Refugees and the 1967 Protocol, which shape informational requirements for recognizing refugee status by identifying the parties responsible for receiving and processing information, as well as integrating that information into decision-making (Ruffer, this volume; Kingston, this volume). The information required for gaining recognition as a refugee, in turn, is influenced by legal protections established in the 1954 Convention Relating to the Status of Stateless Persons and the 1961 Convention on the Reduction of Statelessness. As noted by Kingston, statelessness has many causes, and foundational is a lack of identity documentation, which in turn may stem from a lack of birth registration. Rights of citizens to identity and birth documentation are established by Article 24 of the 1966 International Covenant on Civil and Political Rights and Article 7 of the 1989 Convention on the Rights of the Child. Again, while these legal instruments do not guarantee implementation of national processes, they do establish international legal norms against which nation-states can be measured and appeals can be made. Further, when nation-states do embark on fulfilling these obligations, those processes can be informed either directly by the law or by de facto standard practices that have emerged.

In addition to the informational aspects of refugee status and identity documentation, informational rights of refugees are also informed by the 1948 Universal Declaration of Human Rights (UDHR), as well as the December 2013 UN General Assembly resolution 68/167, which voiced concerns over the impact of surveillance on human rights (Kingston, this volume). These laws not only establish the right to nationality, they also establish and then further draw attention to a right to privacy, which increasingly is being used to shape data protection laws and data management policies.

Data Management

The roots of data management emerge from policies of data availability, affecting producers and consumers alike.

For producers, recent calls for open data, particularly in support of open government, have been made at the UN through its Division of Public Administration and Development Management. As opposed to a purely UN-driven effort, international coordination on open data has been taken up by the Open Government Partnership (OGP). This multi-stakeholder body has seen significant growth. Launched in 2011 on the sidelines of a UN General Assembly Meeting, it has quickly grown from 8 to 70 participating nations, together with their civil society organizations.

It is important to note that while international policies may be directed primarily toward national governments, everyone with a mobile phone, including forced migrants transiting national borders, have become data producers. The treatment of these boundary-spanning data and their producers calls for greater international, regional, and multi-lateral policymaking and implementation, as presaged by Israeli scholar Tene's (2013) "Privacy Law's Mid-Life Crisis: A Critical Assessment of the Second Wave of Global Privacy Laws." The implementation of the EU data protection law in 2018 may serve as a model or become a de facto standard for international data management. For refugees, the desired protections may vary by individual or circumstance. On the one hand, data privacy protections could create physical protection where parties to a conflict are unable to track the routes of fleeing civilians; on the other hand, refugees on a dangerous route (e.g., a desert or sea) may want to be discovered and rescued.

For data consumers, including Tomaszewski (this volume), and Martin and Singh (this volume), data availability is crucial both for spatial analyses and the effectiveness of GIS at all levels, as well as to unlock the potential of so-called big data for macro-level modeling of forced migration. In both domains, open data policies and practices of international bodies can not only provide critical data, but can establish norms for national governments.

While the move toward open government and in some cases e-government will make more data available, reviewing a report titled "Aligning Supply and Demand for Better Governance: Open Data in the Open Government Partnership" (Khan & Foti, 2015b) suggests it is unclear whether the data needs of forced migration researchers will be fulfilled by open government initiatives. Data sets providing insight into government operations and decision-making that are highly valued for open government may hold little value to this research community. On the other hand, global coordination on data releases is likely to benefit researchers both in establishing norms around releases and also creating the potential for greater standardization on measures. The report includes analyses of the prevalence of such topics as machine readability, data standards, and frequency of update in member countries' open data plans. These are positive signs for data scientists, but much work will be needed for their implementation.

Jurisdiction 2: National

Whether developed within the context of international frameworks or independent of them, national-level policies are where direct impacts can be most easily observed. The organizations defining this jurisdiction are more diverse and more directly reflect the domains of interest here. These include UN specialized agencies active in the provision of services to the displaced, multilateral agencies, including regional governments, and national legislative bodies such as parliaments, that establish the fundamental laws of network technologies, refugee rights, and data management.

Network Access

At the national level, access to fixed and wireless telecommunications services, upon which humanitarian organizations and the displaced themselves rely, is shaped by a multiplicity of telecommunication policies that affect everything from spectrum use to market structures, availability of handsets, and even prices. Three critical aspects of the forced migrant condition shaping their network access include: (1) inherent mobility, which makes them more likely to be affected by policies for cellular infrastructure; (2) political sensitivity to any perception of permanence, which can generate informal government intervention; and, in some cases, (3) proximity to armed conflict, which in turn generates national security concerns and potentially surveillance and censorship.

As a result of their mobility, as noted by Schmitt et al. (this volume), forced migrants are affected by national spectrum management laws and policies, which shape, and some argue limit, the range of deployed technologies. Policies promoting experimentation, and, in turn, generating more modular, rapidly deployable network access are critical to ensuring continuous connections for the displaced, particularly while on the move. In addition to spectrum policies, those influencing regulatory governance will also have indirect effects on whether and how "refugee friendly" technologies come to market. For example, policies promoting regulatory independence can be critical to new technology diffusion (Yates, Gulati, & Weiss, 2013).

Forced migrants are also affected by policies due to their lack of permanence. While their mobility makes cellular networks the most important infrastructure, the backhaul provided by fixed networks is also critical. However, build out of fixed networks to serve refugee communities, particularly in camps, can be problematic.

Whether fixed or mobile, the build out of infrastructure for temporary settlements must consider the established levels of connectivity in the local host community. Any extension of service to settlements must ensure equal access on equal terms. This issue was significant in the rollout of infrastructure and the related development of online educational programs in Dadaab Camp, Kenya. While the original intention was to develop the infrastructure and programs to support higher education for refugees in the camp, in the end the education center was located outside the camp, providing service to both camp residents and the local community. While this likely benefitted camp/community relations, the distance to reach the facility was challenging for some refugees.[3]

Also related to permanency of refugees is the permanency of infrastructures. While mobile and wireless infrastructures can appear to be less permanent than fixed infrastructure, they still require a certain level of planning as well as investment. Mobile carriers may be hesitant to invest in infrastructure where demand is perceived as temporary or at least uncertain. Even in situations where conflicts are protracted, with a high level of certainty refugees will not be able to return home, host government policies could introduce an air of uncertainty. For example, the Government of Kenya's announced plans to forcibly return hundreds of thousands of residents from Dadaab back to Somalia creates uncertainty, which has been shown to discourage investment (d'Halluin, Forsyth, & Vetzal, 2002; Maitland, Bauer, & Westerveld, 2002). Further, even where the government actions do not negatively affect market estimations, government fears that an aura of permanency will exacerbate community/refugee tensions, or a political backlash against those who agreed to host the refugee population, may lead them to discourage or in fact prohibit infrastructure build out.

The third dimension of refugee status affecting network access and policies is their proximity to conflict. National governments are frequently leery of displaced persons fleeing armed conflicts as they can bring with them elements of insecurity. For example, former fighters and military deserters, potential recruits (typically young men), and politically active persons with strong convictions and ties to parties to the conflict may exist within the general displaced population. Network access, particularly access to the Internet and social media platforms, is seen as a potential avenue for coordinating military action back home. Also, where armed conflict has a broader regional or global agenda, as is the case with al-Qaeda, some fear access may enable recruitment for terrorist activities to be conducted in the host community. Ostensibly, this fear was partially responsible for the Kenyan government's repeated calls in 2015 and 2016 to disband Dadaab camp (Human Rights Watch, 2016).

Further, in January 2016 the data services offered by three mobile carriers in Za'atari camp were "throttled" by the Jordanian government, meaning they directed the mobile carriers to limit data service speeds. UNCHR reported the service degradation as due to "security concerns." The agency has been unable to take definitive action to restore service to the refugees, however it did push back on these service limitations, noting the benefits of a well-informed refugee population (UNHCR, 2016).

These situations are complex, and policy responses need to balance concerns of a number of stakeholders. In the case of Za'atari, refugees enjoyed unfettered, although low quality and likely surveilled, access to voice and data services from nearly the time of the camp's opening in July 2012 to January 2016. As documented in several reports and articles (see Maitland et al., 2015; Maitland & Xu, 2015; Xu & Maitland, 2016; Schmitt et al., 2016), the refugees came to depend on data services for a variety of communication and information services. These connections not only provided critical information about loved ones, but also, through access to social media and YouTube, helped combat the boredom of life in exile and without work.

At the outset, the throttling of access came with little justification. As I witnessed in March 2016, refugees were rightfully clamoring for explanations about the lack of data services through community meetings with camp leaders. The refugees were facing a true hardship, and unfortunately the managers did not have answers, at least not answers they could share. Additionally, as the camp is located relatively close to an urban center of 80,000 Jordanians, it was soon clear that the throttling was targeted exclusively at the camp.

While the Jordanian government may have had well-founded security concerns about the camp, this example points to several shortcomings in the implementation of their security strategy. In particular, it highlights the need for well-established policies and practices for balancing security and connectivity needs. Information rights principles, such as a proportionate response,[4] promote targeting problems while protecting the rights of the majority. Also, the principle of transparency in actions, if applied, would have generated reasoned explanations, in turn communicating respect and understanding of the hardship.

That being said, the actions of the Jordanian government are in line with those of many other governments. The ability to enact programs of surveillance and censorship, including preventing or limiting network access altogether, are expanding under umbrella policies for national security. As noted by Human Rights Watch (2016) and various scholars, post 9/11 and even more recently, terrorist attacks are often followed by special provisions to curb information access rights, highlighting the tension between information rights and national security (Castells, 2015; Deibert, 2015; Goodman, 2016). Since some refugee-generating conflicts have terrorist dimensions, as the case of Jordan suggests, refugee network access risks being influenced by these changes. They will also have implications for data management, as will be discussed subsequently.

Refugee Rights

At the national level, legislative bodies shape the information needs of refugees through two mechanisms. First are laws determining rights and responsibilities in refugee status determination (RSD) processes, and second are laws influencing access to rights granted refugees. These processes and rights can vary and change over time.

These changes and the link between multilateral and national policy and programs are demonstrated in the case of Ecuador's adoption of the regional Cartagena Declaration. As described by Esthimer (2016), Ecuador is an exemplar in refugee rights in the region, having incorporated the Cartagena Declaration refugee definition into domestic law in 1987. In 2008,

Ecuador took an enormous step in its commitment to refugees by incorporating in its new constitution the "principle of universal citizenship." The provision removed entry visa requirements, and recognized mobility, citizenship, and the ability to safely seek asylum as fundamental human rights, regardless of migration status or nationality. While the policy was subsequently adjusted, it provides an example of how national policy affects refugees' information needs and requirements. With Ecuador having eliminated any visa requirement, the informational needs of refugees fleeing to that country are significantly reduced, both in the information required to properly and legally enter the country, as well as the information subsequently needed for immigration processes.

Similarly, in response to the Syrian crisis, Brazil implemented a special humanitarian visa policy, which also adjusted the information and documentation requirements, particularly as humanitarian visa holders have the right to apply for refugee status. Visas have been granted to 7,380 Syrians with 2,000 applying for and being granted refugee status. However, as noted by Human Rights Watch, this program has fallen short of the durable solution true resettlement offers.[5]

Data Management

The management and protection of forced migrants' data are influenced not only by international legal regimes, but also by national data protection policies. For example, in the U.S. and EU, data protections related to refugees are informed by, but not defined by, national data protection policies. That process of influence is often established in courts where refugee advocates seek clarification as to whom protections apply. Often, as non-citizens, and due to national security concerns, data protections afforded citizens are not available to asylum seekers.

Within the national sphere, and of particular interest for refugees, are identity documentation policies. While international law establishes the basis for these rights, they are based upon a number of assumptions that in practice, as observed in national identity systems, expose the need for more granular conceptualizations. In an essay titled "Identity Systems Don't Always Serve the Bottom 40%," which articulates insights from the World Bank's Identification for Development Program, Brewer, Menzies, and Schott (2015) call for further disambiguation of "identity," "registration," and "documentation" in all contexts. While their analysis focused on the broader domain of international development, its insights are critical for the study of data policies for refugees.

Finally, at the national level, refugee research benefits from national open data policies. The complexity of the environment within which national open data policies are and will continue to be implemented is daunting. In their research to define a process for planning and developing open government data programs, Dawes, Vidiasova, and Parkhimovich (2016) retreat to a comparative analysis of just two cities, New York and St. Petersburg, Russia, to test their model. Even at that lower level, the complexity is a challenge. However, efforts of the multi-lateral Open Government Partnership to facilitate sharing of best-practices between governments may help speed up implementations (Khan & Foti, 2015a; Khan & Foti, 2015b).

Jurisdiction 3: Organizational

Actors involved at the organizational level are numerous. These include specialized UN agencies and their NGO partners, infrastructure providers, and government organizations. Here the term infrastructure providers is used to denote a broad range of organizations, including those providing publicly accessible infrastructure (the traditional telcos and ISPs), as well as humanitarian service providers, including new entrants, such as the Dutch effort RefugeeHotspot, who provide WiFi access to refugees in their offices, community centers, and schools in the Netherlands. Government organizations are those interfacing with refugees, such as Ministries of Health and Education, and the police, which implement policies and define practices, sometimes of their own accord, and in other cases to maintain compliance with national policies.

Network Access

While historically network access was influenced primarily by national government policies directed at licensed fixed and mobile carriers, for refugee services organizational policies affect network access in a number of ways. For example, as discussed in Schmitt et al., (2016) provisioning access, particularly where bandwidth is limited, often prioritizes humanitarian organizations' over displaced persons' needs. As camps are set up, network access is first secured for the agencies providing services. Over time, discussions with public network operators may ensue, or efforts of others such as donors or technology firms may improve connectivity for both communities.[6]

In some cases, the terms and conditions by which refugees access services are the result of collaboration between network and humanitarian service providers. For example, in Za'atari Camp, UNHCR distributed mobile SIM cards from one carrier, Zain. In Rwanda, WFP and UNHCR worked with the mobile carrier AirTel to provide mobile money to Congolese refugees (Tafere, Katkiwirize, Kamau, & Nsabimana, 2014). These programs and their implications for network access were shaped by organizational policies that justified the provision of network access in service of the broader goals of UN-to-refugee communication and food distribution, respectively.

Policies such as UNHCR's prohibition on charging for services affect the ways in which network access and the related demands for electricity are fulfilled. For example, the prohibition on charging for services means either equal service must be available to all at no charge, or humanitarian organizations must establish a process for providing subsidies or service to individual beneficiaries. In their report Connecting Refugees, UNHCR (2016) calls on carriers to make special lower-cost plans for refugees. They also call for plans to provide subsidies such that the most vulnerable persons of concern can maintain network access. Additionally, the report recognizes the benefits of multiple modes of access and calls for the use of WiFi, providing free access through community centers. In short, previously UNHCR's policy of not charging for services and viewing network access as discretionary have hindered refugee access. However, policy changes are afoot, recognizing the importance of connectivity to refugees and opening possibilities for UNHCR to mainstream network access as a program.

Where access to backbone infrastructure is available, a plethora of refugee connectivity efforts have sprung up, particularly across Europe. In Croatia, two enterprising small firms, Open Net and Mesh Point, bucked government policy to provide WiFi access.[7] The aforementioned Dutch effort RefugeeHotspot, an initiative of their Internet Society, connects refugees at various asylum and community centers. Also, the Disaster Tech Lab, an Irish initiative with a global and disaster-focused mission, has expanded access to refugees in Greece.[8] Finally, Maitland and Bharania (2017) discuss the role of interorganizational relations between NetHope, Cisco, and Google in deploying networks in over 40 camps throughout Greece and nearby countries. These initiatives vary in both their sustainability and scalability, but nevertheless present important new ideas and innovations to the refugee community. As such, it will be interesting to watch the evolution of their policies concerning network access as they expand the number and diversity of their implementations.

Refugee Rights

At the national level, informational components of refugee rights are reflected in national RSD procedures as well as the UNHCR *Handbook and Guidelines on Procedures and Criteria for Determining Refugee Status under the 1951 Refugee Convention and the 1967 Protocol relating to the Status of Refugee,* frequently referred to as the UNHCR RSD Handbook. Clearly, where UNHCR is responsible for RSD the policies and practices adopted in that realm will be influenced, if not dictated, by UNHCR data protection and other policies.

While refugee rights are specified in legislation made in national parliaments, it is actually lower-level government organizations that devise policies and practices directly affecting information needs of displaced persons engaged in the RSD process. In a study of RSD in Albania, Dyduch (2016) notes the changing nature and inconsistent application of that country's policies in cases heard by the government's Directorate for Nationality and Refugees (DfNR) between 2006 and 2011. Informational components of the process, such as the right to be informed about the process, the nature of the evidence used to determine status, and the authority's validation of the information provided, varied across cases and over time.

These informational aspects of RSD are complicated, with applicants presenting documents and evidence, and staff assessing their credibility. In the UK, according to their 2015 manual titled "Asylum Policy Instruction: Assessing Credibility and Refugee Status," documents and information presented by applicants can come from a variety of sources. Not only are applicants encouraged to provide information from websites and online databases, similarly, RSD staff are encouraged to assess the validity of information and documents presented to them using data from UNHCR's refworld.org and other web portals. Further, it is clear that as greater volumes of multimedia information are posted online, establishing the credibility of claims will become increasingly difficult. This is particularly true for applicants having access to much greater volumes of data with which they might construct *invalid* claims, which are undetectable by staff who have access to the same data. Imagining widespread access to YouTube, the likelihood of this looming (if not already existing) conundrum is exemplified by a passage from the manual that states: "for example, where the *published record* [emphasis added] of what occurred at a particular demonstration clearly does not coincide with the claimant's account and their claim to have been present or victimized is not credible, the material fact should be rejected" (UK Home Office, 2015, p. 16).

Yet, at the same time, some countries are placing less emphasis on evidence and its credibility. As noted in the aforementioned Esthimer (2016) Ecuadoran case, the national policy changes in turn spurred revisions in operational policies. Aligned with its constitutional changes in 2008, the country also implemented a system called the Enhanced Registration Project. The system streamlined the refugee registration process by shortening the waiting period for an asylum decision from months to one day. In addition to streamlining registration, it also made registration available in remote regions. Mobile teams of government workers brought registration directly to refugees along the northern border with Colombia. Many likely had never had access to the asylum system. It is likely the streamlined registration made the mobile teams feasible, thereby not only reducing information requirements for both the government and refugees alike, but also expanding services to those in need.

Informational components of refugee rights go beyond RSD and include a number of issues related to life in asylum and differences between refugees and citizens. Special forms of identification, and special informational and identification requirements to access government services, once persons are granted asylum, may exist. For example, Landau (2006) describes in the South African context the extreme efforts needed to obtain an identity card and its previously flimsy nature, a paper document, which made it non-durable in normal circumstances and subject to destruction by corrupt police (p. 317–319).

Similarly, informational requirements for resettlement can also be extensive. Depending on the country, resettlement may require far more extensive identification information and circumstance validation data (health, disability, familial relations, etc.) than mere refugee status determination alone. For refugees nominated for resettlement in the U.S., applicants submit extensive information, including biographic and biometric data, which are then cross-referenced against a variety of databases. Examples of the latter include the Department of State's Consular Lookout and Support System (CLASS), which itself draws data from a variety of sources, the Department of Homeland Security's Automated Biometric Identification System (IDENT), and the Department of Defense Forensics and Biometrics Agency (DFBA)'s Automated Biometric Identification System (ABIS).[9]

Further, as discussed by both Kingston (chapter 3) and Maitland (chapters 4 and 7), technology use, which, given the autonomous nature of many field offices, frequently diffuses in a bottom-up manner, and can be the source of policies that trickle up through an organization. Hence, information policies related to refugee rights may have their impetus either in actual technical implementations, as is the case for biometrics, or may reflect more top-down policy prescriptions.

Data Management

The organizational dimensions of data management policies are critical. While international and national bodies may lay out general frameworks, the nuts and bolts of data management processes and practices are established in organizational policies. As reflected in chapter 7, organizations are driven to develop policy due to their control over the processes by which data are captured, stored, managed, and shared.

The UNHCR (2015) Policy on the Protections of Personal Data of Persons of Concern to UNHCR provides a comprehensive approach, with a few limitations, to data protection. The policy is designed to apply to refugees' and IDPs' personal data only, often referred to as personally identifiable information (PII). The handling of aggregated or anonymized data is regulated by a different policy, UNHCR's Information Classification, Handling, and Disclosure Policy. However, the two can be related. Data scientists have shown it is possible to derive identities from analyses of large volumes of aggregated and anonymized data.[10] In addition to bounding the types of data, the policy also specifies the duration of protection. In particular, it states the policy continues to apply even after "persons are no longer of concern to UNHCR" (p. 8).

The policy is less concerned with operational guidance, suggesting that will be offered elsewhere, but instead lays out a system of governance of data protection, including identifying positions responsible for data protection—the Data Protection Officer in the Division of International Protection, and the Data Controller, as well as the Data Protection Focal

Point. The Data Controller is responsible for policy implementation and compliance, and the Focal Point is the highest position among the protection staff in an office/operation.

While steering clear of specific operational guidance, the policy does pay significant attention to data sharing and interactions beyond UNHCR's organizational borders. It lays out stipulations and scope for sharing, indicating the policies also apply to implementing partners. The policy also stipulates the conditions under which data are to be shared with national authorities, particularly those involved in law enforcement. Additionally, the policy devotes space to the practicalities of rights of refugees, particularly for requesting correction or deletion of data. Detailed procedures for lodging complaints and required periods for responding are identified.

Beck and Kuner (2015), a team consisting of a UNHCR staff member and a legal scholar, provide background to and a discussion of the implications of the policy. They note the policy's concurrence with and adoption of evolving principles reflected in several UN and multilateral instruments related to data privacy, in particular the Madrid Resolution.[11] They also praise the document's focus on individual rights, in this case those of refugees.

As data protection and information policies and practices continue to evolve and disseminate throughout UNHCR, they are doing so with increasing transparency to refugees and the public. For example, publicly available documents include a model data-sharing agreement for data sharing between UNHCR and governments[12] and a 2015 Q&A document apparently released to address concerns of refugees relating to the limits of data sharing with the Government of Lebanon.[13] Our own projects and data-sharing agreements with UNHCR in 2015 and 2017 suggest a growing sophistication in refugee data protection. While our first agreement was a mere two pages in length, the second had extended to seven pages, with a much more granular specification of data use and security requirements.

These efforts of transparency and further protection of the rights of refugees, including provisions in the data protection policy, are kept on international agendas in part by the work of independent voices. For example, organizations such as Privacy International[14] are working to analyze and promote refugee privacy rights. Also, independent critical analyses by scholars, such as those offered by Jacobsen (2015) in her book, *The Politics of Humanitarian Technology: Good Intentions, Unintended Consequences, and Insecurity*, can play an important role in shaping the discourse of refugee rights. In particular, her analysis of UNHCR's use of iris scanning technologies identifies both risks and benefits and the potential for subsequent harms. In articulating these harms, she introduces the notion of the "digital refugee," for which policies and practices must also provide protection. This concept is taken up in greater detail in chapter 11 of this volume.

As technologies evolve and create greater opportunities for data sharing (as discussed in chapter 7), the need for interorganizational policymaking and enforcement will be critical. As noted in chapter 4, where data flows are related to monitoring and evaluation, contractual mechanisms can provide important safeguards for data quality, transmission, and security. Otherwise, protections for refugees' personally identifiable information typically are stipulated in project MOAs or the non-disclosure agreements referred to above.

The potential of widespread and almost uncontrollable data flows, in which refugee rights to their own data are difficult to protect, are presaged by an analysis of the U.S. system. Kalhan (2013) raises the question of "whose data" in his analysis of what he calls "automated immigrant policing." In particular, the analysis points to the U.S. federal government Secure Communities program, in which databases of Department of Homeland Security (DHS) were integrated with those of the FBI and Department of Justice (DoJ). The resulting system was designed to automatically identify the immigration status of persons accused of crimes. The analysis demonstrates the wide ranging forums in which issues of data ownership, in this case of fingerprint data, are being debated, and sheds light on the organizational challenges to maintaining individuals' rights to access data across their lifetimes.

Another sticky area for organizations is data collection, in particular informed consent. While the UNHCR (2015) policy requires informed consent in the collection of all forms of data, whether for operational or research purposes, valid questions remain as the extent to which the power relationship, and the potential for a lack of technological sophistication on the part of some refugees, can make true informed consent impossible. Answers to these questions may be found, at least to some extent, in the extensive legal scholarship on privacy, particularly those studies focused on informed consent. For example, Bechmann (2014) provides a critical analysis of consent, actually referring to "non-informed consent cultures" in an analysis of Facebook and other online consent processes.

Jurisdiction 4: Individual

The last jurisdiction level is individual data ownership, where the displaced persons themselves can be seen as developing and differentially enforcing personal data management policies. While it can be argued that personal data management policies are broader than simply privacy, much attention has been paid to the challenges of privacy self-management, particularly in the context of U.S. jurisprudence (Solove, 2013). Acquisti, Brandimarte, and Loewenstein (2015) delineate three streams of privacy research that help to explain the challenges people face in personal privacy management. The three domains include: (1) people's uncertainty over the nature of privacy trade-offs and people's preferences over them; (2) the context-dependence of privacy preferences; and (3) the malleability of privacy preferences, particularly in the face of those possessing greater insight into their determinants. In the domain of refugee privacy protections, where power differentials are significant, Aquisiti et al.'s recognition of the change in power that results simply from the transfer of data itself is important. In the refugee services scenario, this translates to yet another power-base for humanitarian organizations that collect significant amounts of personally identifiable information.

Collection of refugees' personal data is driven in part by accountability concerns. Unlike normal citizen/government relations, there is a high level of justification expected for every expenditure. These justifications and accounting of monies spent require extensive data collection. However, these processes are not unique to refugee affairs, but instead are endemic in humanitarian affairs and international development (Madianou, Ong, Longboan, & Cornelio, 2016; Tapia & Maitland, 2009).

While less researched, individual policies concerning security are also likely to play an important role for refugees of the future. While many researchers of individual security focus on providing enhanced security technologies, a few have waded into the area of security practices (e.g., Egelman et al., 2014; Harbach, De Luca, Malkin, & Egelman, 2016). As the field evolves, it is likely to separate out policies and behaviors, viewing behaviors as simply the effective implementation of policy. In an interesting study of mobile phone locking behaviors (self-reports), Harbach et al. (2016) examined international differences in locking behaviors in eight countries: Australia, Canada, Germany, Italy, Japan, Netherlands, the UK, and the US. They found significant differences between the countries, with respondents from the US much less likely to use a lock, compared with all others. Also, women and older persons were less likely to use locks. As refugees' use of mobile phones continues to grow, the importance of these personal security policies, and potential differences in sub-populations may emerge.

Individual data management policies for refugees may be even more important than for the general population.[15] It is likely that as digitization continues, those in conflict zones will become more reliant on mobile phones or other mobile computing devices for storage of critical documents due to their portability. Also, while everyone will have to contend with how to be able to read a digital birth certificate 40 years after its creation, for refugees, who no longer have a relationship with their nation state, requesting backup copies or accessing the tools to read an outdated file may be impossible. For these reasons, research on the individual data management policies of refugees is needed.

An Information Policy Research Agenda

As suggested by the discussion above, there are numerous jurisdictions and domains of information policymaking that influence the ways in which the displaced, humanitarian organizations and even researchers can make more effective and beneficial use of ICTs. To ensure research flowing from this agenda is forward-looking and broadly relevant, the focus is on dynamic interactions between the various domains. In constructing these dynamic interactions the goal is to map a cause and effect. The three themes are: (1) dynamic aspects of the relationship between network access and data management policies; (2) dynamic aspects of refugee rights and data management policies; and (3) dynamic aspects of general technological change and data management policies.

Dynamic Aspects of the Relationship Between Network Access and Data Management Policies

As the displaced and humanitarian organizations gain higher-speed access to networks, the types and volumes of data flowing through these networks will expand exponentially. Information policy research needs to understand how data management policies can be developed and enforced as network architectures and access changes. UNHCR's (2016) Connecting Refugees report highlights differences in connectivity speeds and types of networks. As refugees and indeed programs that were served only by 2G voice service evolve to data, how do those refugees and operations begin to implement data policies to protect refugees? Hence, the dynamic nature of a refugee service system staffed by UNHCR and its implementing partners needs to be taken into account in studying data protection policies.

The Syrian refugee crisis, as one involving persons fleeing a middle-income country and arriving in countries with robust network infrastructure (Turkey, Lebanon, and Jordan), can be seen as a bellwether of these changes. Questions that arise include: How will data management policies change as refugee data storage (on smartphones) increases? How will refugee data—as collected, curated, stored, and shared by them—be shared, validated, and integrated into refugee service provision systems? As refugees gain skills in collecting data via mobile devices using tools such as Open Data Kit, as in our Za'atari Camp asset mapping project,[16] will these data simply create another silo? Or will they be recognized as perhaps more valid reflections of refugee thoughts, interests, and priorities?

Dynamic Aspects of Refugee Rights and Data Management Policies

New ICTs have implications for the nature of displaced persons' rights. Yet at the same time, these technical changes raise issues with direct impacts on data management. Despite the circumstance that refugee status implies greater and greater data collection activities on the part of humanitarian organizations, refugee rights require protection of those data. How will this effectively be managed across organizations and, perhaps more importantly, over time? While this question targets the refugee/service provider relationship, the refugee/state relationship is important as well. Often, refugees experience a break in, or indeed a lack of, the state/citizen relationship. Yet the assumption of that relationship subtly underpins much of the data protection discourse. As such, how can data protections follow forced migrants as they cross national borders? Recognizing the changing state/citizen relations as they apply to data protections for forced migrants will be important for developing these policies.

Finally, as alluded to by a recognition of the change of protections that accompany a changing relationship with the nation-state, one must also expect changes in surveillance and censorship. Green and Lockley's (2013) analysis of cyberattacks against Karen refugees in the UK, as well as other examples discussed by Maitland and Bharania (2017), present a set of circumstances in which the persecution by the state reaches across international borders. The issues raised by this action are numerous. First, and simply, refugees who experienced censorship and surveillance in their home country may bring with them a significant mistrust of the Internet, social media, and potentially also digital devices of any kind. This could have significant impacts on humanitarian organizations' abilities to provide protection and limit harm while at the same time expanding the digital basis of their programs. Second, fundamental notions of protection of the "digital refugee" must be considered. How far will those protections extend? Does host country surveillance imply harm to the digital refugee? Finally, in extreme cases, such as that described by Green and Lockley (2013), to what lengths will they go and what forms of protection will the host country, UNHCR, and/or other members of the humanitarian community provide?

Dynamic Aspects of Technological Change and Data Management Policies

In this final section of the information policy research agenda, the focus is on longer-term technological changes and their implications for refugee data management policies. As noted by Tene (2013) in his analysis of global privacy law, three technological forces changing the nature of privacy include big data and analytics, which enable identification of derived relationships, social networking as data flow facilitators, and the migration of personal data processing to the cloud. These trends are likely to have positive as well as negative consequences for the displaced, humanitarian organizations, and researchers alike. For this last group, the effects, and indeed benefits, of big data are exemplified in the work of Bharti, Lu, Bengtsson, Wetter, and Tatem (2015), who used anonymized call records from a cellular carrier together with satellite data, two data sets of different granularities, in their predictions of forced migration in West Africa. While beneficial for researchers, the explosion of data and the ability to potentially derive identities and relationships has implications for refugees' as well as humanitarian organizations' information policies. Those policies must consider how data, particularly those released to the public, could potentially pose harms to refugees. This requires staying up to date on the power of new analytic techniques and machine learning algorithms that might be applied to large datasets managed by humanitarian organizations.

While social networking platforms are anticipated to contribute significantly to big data and have already sparked a wave of research on the role of social media in displacement, the role of cloud storage has not. For example, might each refugee be provided a small amount of cloud storage space in which to store and manage the data they bring with them to the crisis, but also copies of data generated while in temporary asylum? Would the provision of data storage become the new equivalent to shelter for personal property? In terms of services provided by humanitarian organizations, what policies must or should be enacted to specify conditions for cloud-based data sharing? The launch of the UNHCR KoBo server has created a centralized location for storing data and could be seen as a precursor to cloud storage. Will UN-wide data protection policies allow cloud storage or will specialized services meeting UN requirements be needed?

As technology continues to change, how will UNHCR and its partners view technology vis-à-vis established service areas such as sanitation, shelter, and food? UNHCR's (2016) networking report suggests there is an inclination to begin facilitating network access. As these intentions evolve, lessons learned from the broadband arena, such as the evolving quality of service standards, may provide insight. Beyond data protection, there are questions of prioritizing systems levels investments, and defining the quality of network service and levels of security necessary to protect the digital refugee.

Conclusion

Information policies play a crucial role in shaping the information environment of ICTs for refugees and the displaced. The analysis presented here necessarily is just a slice of this rich and important area for research. Taking a pragmatic approach, the scope was reduced through an inductive analysis of this volume's chapters, identifying three domains, namely, networks, refugee rights, and data management. Providing just one or two examples, these domains were analyzed across international, national, organizational, and individual jurisdictions. The resulting survey of network's, right's, and data's implications for information policies considers technologies with different innovation processes. These technologies' impacts were analyzed at different levels, highlighting their interdependencies. At the same time, trends in refugee rights at these same levels were assessed.

Following this rather extensive background, a research agenda focusing on dynamic elements and relationships between cause and effect was proposed. Each element makes a contribution to this volume by highlighting the relationships between network access and data management, refugee rights and data management, as well as general technological change and data management policies. This research agenda can be taken up by scholars in fields as diverse as refugee studies and information sciences. Nearly all aspects would benefit from interdisciplinary collaboration.

Despite its breadth, this information policy research agenda has several limitations. First, it is incomplete in that it did not address issues across all phases of the refugee life cycle. As is the case in other areas of this volume, less attention is paid to fleeing as well as to repatriation. However, the research topics discussed above easily can be extended further into these dimensions and indeed will provide valuable insights to such endeavors. Second, due to its breadth, it ignores the many circumstantial elements mediating relationships between the variables of interest such as refugee rights and data management. Comparative research will be necessary to uncover the commonalities. This is simply yet another factor calling for extensive research into information policies for the displaced.

Notes

1. As reported by the ITU and World Bank's InfoDev ICT Regulation Toolkit resource, http://www.ictregulationtoolkit.org.

2. For a detailed discussion of policy harmonization from the U.S. perspective, see the National Telecommunication and Information Administration's 2008 report "Spectrum Policy for the 21st Century." https://www.ntia.doc.gov/files/ntia/publications/international_spectrum_policy_improvements_report3-13-08_final.pdf.

3. See https://www.usaid.gov/news-information/frontlines/energy-infrastructure/dadaabnet-wiring-worlds-largest-refugee-camp and http://nethope.org/project/dadaab-refugee-camp.

4. https://necessaryandproportionate.org/about.

5. https://www.hrw.org/news/2016/09/19/brazils-chance-leadership-end-refugee-crisis.

6. https://www.usaid.gov/news-information/frontlines/energy-infrastructure/dadaabnet-wiring-worlds-largest-refugee-camp.

7. https://citiesintransition.eu/cityreport/wi-fi-for-refugees.

8. http://disastertechlab.org/2016/08/10/extending-the-lesvos-network.

9. https://www.uscis.gov/refugeescreening.

10. In a sub-field of privacy scholarship referred to as de-anonymization, researchers investigate the bounds of different types of data sets for the potential to derive identities. One of the earliest works in this domain is that of Ramakrishnan et al. (2001) in their article titled "Privacy Risks in Recommender Systems" (published in IEEE Internet Computing).

11. http://privacyconference2011.org/htmls/adoptedResolutions/2009_Madrid/2009_M1.pdf.

12. http://www.unhcr.org/50a646c79.pdf.

13. https://www.refugees-lebanon.org/en/news/44/qa-on-the-data-sharing-agreement-between-government-of-lebanon-and-unhcr-on-basic-information-about-syrian-refugees.

14. https://www.privacyinternational.org/node/300.

15. https://thedigitalresponder.wordpress.com/2014/11/22/a-cybersecurity-wake-up-call-for-emergency-managers.

16. http://cmaitland.ist.psu.edu.

References

Abbott, K. W., Snidal, D., & Abbott, K. W. (1998). Why States Act through Formal International Organizations. *Journal of Conflict Resolution, 42*(1), 3–32.

Acquisti, A., Brandimarte, L., & Loewenstein, G. (2015). Privacy and Human Behavior in the Age of Information. *Science, 347*(6221), 509–514.

Bauer, J. M. (2014). Platforms, Systems Competition, and Innovation: Reassessing the Foundations of Communications Policy. *Telecommunications Policy, 38*(8), 662–673. doi:10.1016/j.telpol.2014.04.008.

Bauer, J. M., & Bohlin, E. (2007). Dynamic Regulation: Conceptual Foundations, Implementation, Effects. In proceedings of the *35th Telecommunications Policy Research Conference.*

Bechmann, A. (2014). Non-Informed Consent Cultures: Privacy Policies and App Contracts on Facebook. *Journal of Media Business Studies, 11*(1), 21–38.

Beck, A., & Kuner, C. (2015). Data Protection in International Organizations and the New UNHCR Data Protection Policy: Light at the End of the Tunnel? Retrieved from http://www.ejiltalk.org/data-protection-in-international-organizations-and-the-new-unhcr-data-protection-policy-light-at-the-end-of-the-tunnel.

Bharti, N., Lu, X., Bengtsson, L., Wetter, E., & Tatem, A. J. (2015). Remotely Measuring Populations During a Crisis by Overlaying Two Data Sources. *International Health, 7*(2), 90–98. doi:10.1093/inthealth/ihv003.

Brewer, M., Menzies, N., & Schott, J. (2015). Identification Systems Don't Always Serve the Bottom 40%. *Just Development Series, 8*, 1–11.

Castells, M. (2015). *Networks of Outrage and Hope: Social Movements in the Internet Age.* Indianapolis, IN: John Wiley & Sons.

d'Halluin, Y., Forsyth, P. A., & Vetzal, K. R. (2002). Managing Capacity for Telecommunications Networks Under Uncertainty. *IEEE/ACM Transactions on Networking, 10*(4), 579–588.

Dawes, S. S., Vidiasova, L., & Parkhimovich, O. (2016). Planning and Designing Open Government Data Programs: An Ecosystem Approach. *Government Information Quarterly, 33*(1), 15–27. Retrieved from doi:10.1016/j.giq.2016.01.003.

Deibert, R. (2015). Cyberspace Under Siege. *Journal of Democracy, 26*(3), 64–78.

Dyduch, X. (2016). Refugee Status Determination (RSD) in Albania. *Forced Migration Review, 51*, 89–91.

Egelman, S., Jain, S., Portnoff, R. S., Liao, K., Consolvo, S., & Wagner, D. (2014). Are You Ready to Lock?: Understanding User Motivations for Smartphone Locking Behaviors. In proceedings of the *2014 ACM SIGSAC Conference on Computer and Communications Security* (pp. 750–761). ACM (Association for Computing Machinery).

Esthimer, M. (2016). Protecting the Forcibly Displaced: Latin America's Evolving Refugee and Asylum Framework. *Migration Policy Institute*. Retrieved from http://www.migrationpolicy.org/article/protecting-forcibly-displaced-latin-americas-evolving-refugee-and-asylum-framework.

Goodman, A. (2016). Blocking Pro-Terrorist Websites: A Balance between Individual Liberty and National Security in France. *Southwestern Journal of International Law, 22*, 209.

Green, G., & Lockley, E. (2013). *Surveillance without Borders. The Case of Karen Refugees in Sheffield. Emerging Trends in ICT Security*. Amsterdam: Elsevier Inc. 10.1016/B978-0-12-411474-6.00032-3.

Harbach, M., De Luca, A., Malkin, N., & Egelman, S. (2016). Keep on Lockin' in the Free World: A Multi-National Comparison of Smartphone Locking. In proceedings of the *2016 CHI Conference on Human Factors in Computing Systems*. ACM (Association for Computing Machinery).

Human Rights Watch. (2016). *World Report*. Retrieved from https://www.hrw.org/sites/default/files/world_report_download/wr2016_web.pdf.

Jacobsen, K. L. (2015). *The Politics of Humanitarian Technology: Good Intentions, Unintended Consequences and Insecurity*. New York: Routledge.

Kalhan, A. (2013). Immigration Policing and Federalism Through the Lens of Technology, Surveillance, and Privacy. *Ohio State Law Journal, 74*(6), 1105–1165.

Khan, S., & Foti, J. (2015a). *2015 Annual Report*. Retrieved from http://www.opengovpartnership.org/sites/default/files/OGPreport2015.pdf.

Khan, S., & Foti, J. (2015b). *Aligning Supply and Demand for Better Governance: Open Data in the Open Government Partnership*. Retrieved from http://www.opendataresearch.org/dl/symposium2015/odrs2015-paper49.pdf.

Landau, L. B. (2006). Protection and Dignity in Johannesburg: Shortcomings of South Africa's Urban Refugee Policy. *Journal of Refugee Studies, 19*(3), 308–327. doi:10.1093/jrs/fel012.

Madianou, M., Ong, J. C., Longboan, L., & Cornelio, J. S. (2016). The Appearance of Accountability: Communication Technologies and Power Asymmetries in Humanitarian Aid and Disaster Recovery. *Journal of Communication, 66*(6), 960–981.

Maitland, C. F., Bauer, J. M., & Westerveld, R. (2002). The European Market for Mobile Data: Evolving Value Chains and Industry Structures. *Telecommunications Policy, 26*(9–10), 485–504. Retrieved from doi:10.1016/S0308-5961(02)00028-9.

Maitland, C. F., & Bharania, R. (2017). Balancing Security and Other Requirements: The Case of the Syrian Refugee Response. TPRC, Arlington, VA September 8–9, 2017.

Maitland, C. F., Tomaszewski, B. T., Belding, E., Fisher, K., Xu, Y., Iland, D., et al. (2015). Youth Mobile Phone and Internet Use in Za'atari Camp, Mafraq, Jordan, Submitted to UNHCR. https://cmaitland.ist.psu.edu/wp-content/uploads/sites/9/2015/01/ZaatariSurveyAnalysis2015November2.pdf.

Maitland, C. F., & Xu, Y. (2015) A Social Informatics Analysis of Refugee Mobile Phone Use: A Case Study of Za'atari Syrian Refugee Camp. TPRC, Arlington, VA, September 25–27th, 2015. https://papers.ssrn.com/sol3/papers.cfm?abstract_id=2588300.

Ramakrishnan, N., Keller, B. J., Mirza, B. J., Grama, A. Y., & Karypis, G. (2001). Privacy Risks in Recommneder Systems. *IEEE Internet Computing*, (5/6): 54–62.

Scheel, S., & Ratfisch, P. (2014). Refugee Protection Meets Migration Management: UNHCR as a Global Police of Populations. *Journal of Ethnic and Migration Studies, 40*(6), 924–941.

Schmitt, P., Iland, D., Belding, E., Tomaszewski, B., Xu, Y., & Maitland, C. F. (2016). Community-Level Access Divides: A Refugee Camp Case Study. In proceedings of the *ACM 2016 Conference on Information and Communication Technologies for Development (ICTD2016)* (p. 11). Ann Arbor, MI. Retrieved from doi:10.1145/2909609.2909668.

Solove, D. J. (2013). Privacy Self-Management and the Consent Dilemma. *Harvard Law Review, 126*(7), 1880–1903.

Tafere, M., Katkiwirize, S., Kamau, E. N., & Nsabimana, J. (2014). Mobile Money Systems for Humanitarian Delivery: World Vision Cash Transfer Project in Gihembe Refugee Camp, Rwanda. In L. R. Vazquez & I. Will (Eds.), *Communications Technology and Humanitarian Delivery: Challenges and Opportunities for Security Risk Management* (pp. 42–44). European Interagency Security Forum (EISF).

Tapia, A., & Maitland, C. F. (2009). Wireless Devices for Humanitarian Data Collection. *Information Communication and Society, 12*(4), 584–604. doi:10.1080/13691180902857637.

Tene, O. (2013). Privacy Law's Midlife Crisis: A Critical Assessment of the Second Wave of Global Privacy Laws. *Ohio State Law Journal, 74*(6), 1217–1261.

UK Home Office. (2015). *Asylum Policy Instruction: Assessing Credibility and Refugee Status*. Retrieved from https://www.gov.uk/government/uploads/system/uploads/attachment_data/file/397778/ASSESSING_CREDIBILITY_AND_REFUGEE_STATUS_V9_0.pdf.

UNHCR. (2015). *Policy on the Protections of Personal Data of Persons of Concern to UNHCR*. Retrieved from http://www.refworld.org/docid/55643c1d4.html.

UNHCR. (2016). *Connecting Refugees: How Internet and Mobile Connectivity can Improve Refugee Well-Being and Transform Humanitarian Actions*. Retrieved from http://www.unhcr.org/5770d43c4.pdf.

Yates, D. J., Gulati, G. J. J., & Weiss, J. W. (2013). Understanding the Impact of Policy, Regulation, and Governance on Mobile Broadband Diffusion. In proceedings of the *46th Hawaii International Conference on System Sciences (HICSS)* (pp. 2852–2861). IEEE. Retrieved from doi:10.1109/HICSS.2013.583.

Xu, Y., & Maitland, C. F. (2016). Communication Behaviors When Displaced: A Case Study of Za'atari Syrian Refugee Camp. (Note) Proceedings of the Eighth International Conference on Information and Communication Technologies and Development (ACM ICTD2016), Ann Arbor, June, 2016.

11 The ICTs and Displacement Research Agenda and Practical Matters

Carleen F. Maitland

Increasing forced migration coupled with continued innovation in information and communication technologies (ICTs) has created a new environment for humanitarian relief. Rife with opportunities, including connecting families, securing identities, accessing education, and developing livelihoods, ICTs ignite a passion for change. Yet, they also generate many uncertainties. This volume has explored these changes and uncertainties, as well as identified their potential and effects, generating recommendations for directions in applied and academic research. Such research can enable a more proactive, as opposed to reactive, approach to technological development, increasing the chances for positive outcomes for displaced persons.

While each chapter has focused on a specific issue or technology, here two cross-cutting themes are presented. The themes arise from the rapid increase in the availability and dissemination of data, together with changes in the broader humanitarian sector, and reflect fundamentally different perspectives on information. The first, the "digital refugee," focuses on data and information as a reflection of individual displaced persons. They are the basis for a digital persona that not only reflects a static picture generated from categorical data, such as gender, nationality, and education level, but also depicts a much richer identity, mirroring and communicating a refugee's lived experience, emotions, preferences, and passions. As in our physical lives, the identity is formed through self-reflection as well as the opinions, and data, of others. As such, it is derived from both the "self and other," where these two elements interact. Sometimes the interaction is mutually reinforcing, while at others it creates conflict.

In contrast, the second theme, "digital humanitarian brokerage," views information as instrumental inputs to and outputs of complex organizational processes. They are manipulated, analyzed, and used, holding little intrinsic value. Instead, value is derived from their role in processes ranging from authentication, forecasting, and decision-making to coordination. These processes, with information as inputs, create value that is poised to become the basis of a new form of humanitarianism. Building upon notions of digital humanitarians (Meier, 2015) and digital humanitarian*ism* (Conneally, 2011), this theme envisions a humanitarian sector in which brokerage becomes a fundamental service, complementing established service provider roles.

In the following pages, I develop these two themes, highlighting their roots in the chapters in this volume and beyond. Next, I discuss limitations and gaps, and then turn attention to practical advice for implementing the research agenda. The chapter closes with conclusions.

The Digital Refugee

The concept of the digital refugee was first posed by Katja Jacobsen in relation to concerns arising from the use of biometrics by aid agencies (Jacobsen, 2015). Issues of privacy and security related to use of these digital identifiers raised questions about the extension of the "protection mandate" to those data. The notion of a digital refugee, being constructed through the processes of and interactions with humanitarian organizations, itself requiring protections, is compelling. However, as this volume's chapters make clear, the digital refugee is being constructed through a plethora of data capture technologies, operated by a wide range of organizations and individuals. Accordingly, the concept is expanded, giving agency to the refugee him- or herself, as a participant in this construction project, and highlighting additional implications of humanitarian organizations' use of technology.

For simplicity, here the digital refugee is framed as a joint construction project. On the one hand, it is an outgrowth of refugee ICT use, with mobile phones and their associated technologies, including cameras and location tagging, an important source. The refugee's role in this project is reflected in this volume's treatment of refugee information worlds (Fisher) and crowdsourcing of maps (Tomaszewski). It is also reflected in discussions of data availability as an input to forecasting (Martin and Singh) as well as in data collected for refugee status determination (Ruffer).

The second parties to the project are the humanitarian organizations and their partners, including nation-states. The systems of nationality, immigration, and social services generate data from and for refugees, creating another side of the digital self. Those systems were deliberated in this volume by Ruffer, Kingston, and Maitland, and rely on the networks discussed by Schmitt et al., in turn providing inputs to the analyses of Martin and Singh.

The digital refugee is a complex, multifaceted construct, potentially analyzed from many angles. Following Jacobsen, the focus here is also protection. International humanitarian law, in differentiating refugees from other types of migrants, creates protection responsibilities borne by host countries and humanitarian organizations. However, as the following discussion makes clear, the basis of data protections in citizenship creates protection challenges for all migrants. The discussion begins with a focus on the digital self as individually constructed, and then turns to the interactions between the self-constructed digital refugee and the humanitarian systems.

Digital Self-Construction

On the individual level, similar to many of us, refugees create and contribute their own data to the building of their digital selves. Oftentimes, but not always, these data are shared via social networks. Everything from geo-tagged mobile phone video evidence of their lived experiences in conflict, flight, and exile, to the data, music, and photos stored on phones, contributes to the construction of this digital self.

Insight into the process of digital self-construction is provided by the rich body of scholarship on ICT use by resettled refugees. The scholarship sheds light on ICT use in relation to both physical life as well as a separate life online. In the former, scholars have found ICT use among resettled refugees contributes to social integration, accessing goods and services, and maintaining ties with distant friends and family (Alam & Imran, 2015). This use has been found to enhance the sense of "being at home" in a new community (Wilding & Gifford, 2013), communicating cultural identity (Andrade & Doolin, 2016), and maintaining family relations (Robertson, Wilding, & Gifford, 2016).

In the exclusively online lives of refugees and forced migrants, online activities take on a heightened importance, providing a freedom and connection difficult to find in the cultural alienation, poverty, and loneliness of displacement. In this realm, Doná (2015) focuses on the role online spaces play in the construction of "home" while displaced. In this state, the materiality of connection devices becomes even more critical, and the devices themselves can come to constitute home. Similarly, Witteborn's (2015) work with asylum seekers in Germany more squarely focuses on life online. She emphasizes virtual practice as a form of "becoming" that is a transformation occurring through processes of self-presentation, co-presence, and political mobilization.

Combined, the insights provided by authors in this volume, together with extant research, suggest future investigations of digital self-construction be delineated in two sub-themes. The first examines the effect of digital self-construction on the lives of individuals and on the refugee community as a whole. Research may seek to understand the implications of how, when, where, and why refugees document their lives, contributing to their digital selves. Increasing use of censorship, by the authoritarian regimes refugees flee, as well as in host countries concerned with security, may affect refugees' use of ICTs differentially. Research is needed to unpack how such experiences affect refugee ICT use in general, and the subsequent implications for digital self-construction in particular. Research should examine how this digital self-construction affects refugees' coping mechanisms, emotional well-being, and resiliency. Further, research might examine how personal maps, used to document a journey or as a reflection of or tool for negotiating or navigating new places in asylum, affect an individual's sense of community. Important differences in use of ICTs and construction of digital selves across various demographics (age, gender, ethnicities) should be examined. Finally, research might explore how these firsthand data, curated by forced migrants themselves, might contribute to the overall data used in digital humanitarianism. For example, could these data be directly crowdsourced to generate more granular and accurate models of migration, rendering predictions not of general populations, but of targeted segments, varying over ethnic groups, gender, space, and time?

A second research sub-theme on digital self-construction focuses on the interaction of this individually constructed self with refugee service systems. Clashes of individual and "official" digital selves are increasing in everyday life. However, where refugees are provided network access through state or NGO programs, and lacking citizenship protections, these clashes may be more significant. Research might address questions, such as: How does the digital representation of "self" interact with a changing refugee status determination (RSD) process? How might geo-tagged video and photographic evidence of lived experiences be used in the RSD process? How will this digital self move across borders in relation to the physical body? Will it cast a "digital shadow?" What security and privacy protections will be granted to the individually constructed digital refugee? For example, will a refugee, who does not enjoy the protections of citizenship, be granted privacy rights to data stored in a cloud service? As a function of their protection mission, will UNHCR offer cloud storage to refugees to house their digital selves? Will refugees trust UNHCR in this role?

Humanitarian System Construction

As a partner to the construction of the digital refugee, the humanitarian system builds a multifaceted persona. This version of the digital refugee has core components, such as those reflected in fundamental or core datasets (for example, proGres). These elements of the digital persona are projected into integrated and inter-connected systems as data are copied and shared. As data age, important distinctions can be drawn between demographic and biometric data less likely to change (sex, digital fingerprints), demographic and biometric data that do change (marital status, weight), and data recording perceptions and experiences, such as that used in monitoring and evaluation programs.

Biometric data are often the lightning rod for privacy concerns, gathering growing unease over the possibility of rampant data collection and sharing. However, greater immediate harms may be caused by disclosure of information concerning sexual and gender-based violence (SGBV), or simply, in the case of a military deserter, the action of registering as a refugee. Hence, protection of the digital refugee must be of broad concern.

Protection of the digital refugee constructed by the humanitarian system must confront several trends presented in these chapters. While the growing volume of data surrounding humanitarian services has been chronicled here and elsewhere, this volume provides unique insights into the humanitarian systems that generate, manage, and share these data. Of note are developments in systems for data sharing at international, regional, national, and local levels. These systems involve not only humanitarian actors, but also private firms, particularly banks. While these firms are entrusted with data by thousands of customers in other realms, it is an open question whether or not refugee data requires special protections.

Also, a related difference in the individually and organizationally constructed versions of the digital refugee is data ownership and control. Questions arise regarding the security, privacy, and control over both forms. For example, when organizations own, control, and manage data collection technologies, including but not limited to biometrics, do individual refugees have a right to, or at least a copy of, those data? If so, what types of technologies facilitate transfer, storage, and secure, long-term management? In the context of service provider networks, how can protection be provided? How will these protection mechanisms differ from those offered for the individually constructed version? Will organizational processes of RSD, resettlement, repatriation, and authentication for services in asylum be biased toward the organizationally versus individually constructed digital refugee?

These questions are just few of the many awaiting the research and program efforts needed to provide answers. This discussion highlights differences and similarities in the notions of the individually and organizationally constructed digital refugee. While both require protections, it is clear the nature of and challenges of the two forms differ.

Digital Humanitarian Brokerage

As a second theme, the concept of digital humanitarian brokerage is reflected in many of the contributions of this volume. The brokerage concept builds upon the observed trends of digital humanitarianism (Conneally, 2011) and digital humanitarians (Meier, 2015). The former focuses on the role of data in humanitarian response, sparked by the relief effort following the Haitian earthquake, with a focus on use by established actors. The latter, coming several years later, articulates a more expansive concept, one in which a new set of actors, primarily the volunteer technical community, confront organizational boundaries of established actors, serving as change agents. The agents have harnessed technological forces, ranging from crowdsourcing, machine learning, and drones, and directed them toward humanitarian efforts. Digital humanitarian brokerage is an extension of digital humanitarianism, highlighting data's role in transforming services, but taking it a step further to transforming the sector. This renewed sector will be better positioned to engage with digital humanitarians as well as more fully and quickly integrate their innovations.

Digital humanitarian brokerage is the transition to a more flexible approach to providing humanitarian services. Where private sector actors, the host community, or the displaced themselves can provide a necessary service, humanitarian organizations will serve as intermediaries, prioritizing needs and distributing funds. Yet, unlike a traditional arms' length broker, the humanitarian community will maintain its responsibility to fulfill its protection mandate. Part of its ongoing value will be monitoring markets and back stopping where failures occur. In some circumstances, brokering may be infeasible. In others, nearly all services will be brokered. Whereas current attention to digital brokerage has focused on data (Hellmann, Maitland, & Tapia, 2016), digital humanitarian brokerage focuses on *services*.

The specific drivers of digital brokerage include: (1) increasing data sharing in the provision of humanitarian services, which underpins the brokerage process; (2) greater connectivity of refugees and the displaced, allowing them to access services as well as be accessed by service providers online; and (3) cultural changes within the sector related to humanitarian reform and the need to be more open and flexible, and recognize refugees not as beneficiaries but as partners. These trends have already laid the groundwork for digital brokerage, with several examples reflected in this volume.

Examples can be drawn from across the refugee lifecycle. As discussed by Ruffer (this volume), information used in refugee status determination (RSD) is brokered by UNHCR to national RSD programs around the world. In this program, UNHCR vets information from a variety of external sources, providing a centralized portal via its refworld.org website. Similarly, in the data sharing described by Maitland (chapter 7), UNHCR's registration database proGres stores refugee ID numbers, which serve to connect data, coordinate services, and account for spending, across organizational networks. UNHCR provides this digital identifier brokerage to the sector. As more nations take on registration responsibilities, this role could change.

The digital humanitarian brokerage trend is most clear in the move to digital cash programming, as discussed by Maitland (chapter 7). The transition from "in kind" aid, meaning the material resources of food, shelter, and clothing, to providing digital cash is occurring through brokering of services. This transition is one example of the aforementioned cultural changes that have been ushered in as part of humanitarian reform. While for decades the inefficiencies of in kind aid were well understood, the risks of fraud and insecurity associated with handing out physical cash were considered too great. However, digital cash, and its ease of distribution, as well as the capability to have beneficiaries themselves transact digitally, has changed the calculus. As noted by Maitland's description of three different digital cash/voucher programs, all require coordination of services involving banks. However, the role of the World Food Program goes beyond coordination, since it has the unique expertise required to monitor the market and step in to provide food if digital service is failing. This expertise and protection mandate is fundamental to digital humanitarian brokerage.

Further developments in brokerage are reflected in an engagement with UNHCR by the author in the form of a university class project, which examined the potential of digital brokerage in worldwide offerings of university scholarships to Syrian refugees. The project had students design the concept for an online system that would allow nations, organizations, and universities offering scholarships to post the offers on a portal. The portal would allow UNHCR staff to vet the offers, ensuring adequate protections in the form of language training, availability of psychosocial support, and a clear indication of the potential to remain in the country following the educational program. Refugees would be directed to the online portal, accessible via mobile phone, to apply for positions. As a broker, UNHCR would educate both scholarship providers on the needs and rights of refugees, while also educating refugees on their rights.

This example highlights factors likely to shape the balance between brokering and direct service provision. In this example, the service, education, is being offered at a distance, with no potential for direct engagement by a humanitarian organization. Hence, UNHCR has the option not to participate or broker. In not participating, scholarship providers and students may find one another directly online, a situation more likely to occur as refugees increase their online presence. In brokering, UNHCR provides its protection expertise, from which the displaced benefit.

In comparison, in the digital cash for food example, food is brokered rather than directly provided. However, there is much greater oversight required in this digital broker role, as compared to the scholarship example. In the digital cash for food programs, not only did WFP "make the market" by vetting food vendors and designating the payment mechanism, they also continually monitor the market's food quality and prices. This enhanced broker role is necessary due to the expertise required, as well as food being a critical commodity. As digital cash and voucher programs expand to additional sectors, facilitated by the "multi-wallet" payment infrastructure, choices will be made about where to broker and where to continue to provide direct services. Each humanitarian organization is likely to develop a continuum of service modes, ranging from fully "direct service provision" to fully "digital humanitarian brokerage"—and with hybrid arrangements, such as the Dadaab case.

Two of the more pure forms of brokerage envisioned in this volume are related to finding locations for temporary asylum and forecasting. Ruffer (chapter 2) discusses the potential for brokerage by UNHCR in matching refugees with countries for temporary asylum or resettlement. In such a system, UNHCR would qualify and register refugees, who would then be entered into a system for global placement. This centralized process would reduce the burden on states nearby conflicts, who now bear a disproportionate share of the global displacement burden, while providing a better match between needs and abilities of nation-states to offer protection.

Similarly, Martin and Singh (chapter 9) foresee an enhanced role for forecasting in the sector. Forecasting, often a value-added function of brokerage, enables bringing together resources and service providers in a timely fashion. Similar to other well-known brokers, such as Amazon.com in e-commerce and stock brokers in financial services, awareness of trends and foresight will be increasingly valued in a humanitarian community and is likely to play an important role in digital humanitarian brokerage.

Brokerage's role and ultimate effect on the humanitarian system will depend on the ability of humanitarian organizations to fully integrate new technologies across organizational boundaries. As highlighted by the examples of three distinct digital cash/voucher programs in chapter 7, these deployments must take into account both constraints and opportunities available in the local context. Research is needed to better understand this balance, and its impacts on service and those being served.

Finally, as programs of digital humanitarian brokerage increase in number, there is a need for scholarship on how the programs affect beneficiaries. Would the displaced be better served by a large number of highly specialized and localized service providers? Or do digital services brokered by centralized entities better promote high-quality service? Further, researchers should examine how digital humanitarian brokerage affects refugees both as a vulnerable population as well as a population embracing and enjoying online life and the agency it provides. As a vulnerable population, how general human rights, rights to security, privacy, and even information itself are affected by digital humanitarian brokerage need to be explored. This research is particularly critical given the lack of citizenship-based protections, which contributes to the power differential resulting from refugees' poverty, trauma, and need. The impact of policies stipulating protection of refugees' data will be determined by how effectively and consistently (both geographically and over time) they are implemented.

On the other hand, refugees' embrace of digital life may serve as a natural platform for digital humanitarian brokerage. As agentic users of technology, research on how they prefer to receive support, and what benefits and harms they foresee in an era of digital support provision is needed. Extant research on digital cash programs suggests broad acceptance but harms as well. Such research can contribute to fundamental scholarship on the role of online life in displacement, generating important insights into the spatial boundary spanning affordances of ICTs. It can also inform program design, defining the elements of context requiring unique approaches and where more scalable solutions are appropriate.

Limitations and Gaps

As the above research agenda could not possibly encompass all the important research ideas reflected throughout this volume, the volume itself is limited in scope. In the following, three general issues underrepresented in this volume are briefly discussed.

Refugee Lifecycle Perspectives

In the introduction, a framing of the refugee lifecycle was introduced. This framing positioned elements of the agenda, helping readers to understand the various contexts of ICT use as they relate to displacement.

As stated previously, the agenda presented here primarily focuses on temporary asylum. As such it deals with issues arising after the transit phase, although this is touched upon by Fisher, but before resettlement or repatriation. The flight to Europe from North Africa and the Middle East is starting to provide empirical evidence to fill the long-standing gap in our knowledge of information needs and ICT use during flight. However, scholarship on transit in other contexts such as sub-Saharan Africa and Asia is also needed, particularly as the effects of context, including infrastructure availability, the role of middle men, and relative wealth and education, need to be defined.

Still, the most under-researched domain in ICT use by the displaced is repatriation. Given the two large repatriation efforts currently under way, namely from Kenya to Somalia and Pakistan to Afghanistan, the potential for fresh insights exists. The latter is a continuation of former repatriation drives, begun roughly in 2002, described through compelling personal stories and with interesting historical context by Moorehead (2006), as well from the perspective of the UK's return program by Whiittaker (2006). Repatriation is very interesting from a data management perspective because it often ends the formal relationship as a "person of concern" between the refugee and UNHCR. Currently, Pakistan Refugee ID cards are being destroyed as refugees repatriate to Afghanistan. It is unclear how former refugees will be received in Afghanistan and the forms of identity documentation they will receive. The nature of the transfer of data from UNHCR to the Afghan government is also an open question. The same question can also be posed in the circumstance of Kenya and Somalia, and may be even more interesting given the challenges of the latter.

Other ICTs, Uses and Issues

As noted in UN OCHA's (2013) Humanitarianism in the Network Age, the new techniques of crowdsourcing and crowdseeding may not only be used to generate data and ideas, but can also be used to vet, process, and analyze them. As noted by Meier (2015), these developments could have important implications for humanitarian organizations. While typically used in acute crises such as natural disasters, they are slowly gaining acceptance in refugee services. UNHCR, through its Innovation Ideas program, has crowdsourced ideas on how to improve refugee access to services.[1] A more recent project involves crowdsourcing innovation via a 3D design lab within the Za'atari camp.[2]

Also, building on previous research on the economic lives of refugees (e.g., K. Jacobsen, 2005), an important area in which ICTs may open great potential is livelihoods. Studies examining the role of online work as a legal mechanism for earning a living in countries where refugees are not allowed to work are needed. One effort launched in Kenya's Dadaab camp by a U.S.

non-profit called Samasource, farmed out online work to camp-based refugees.[3] The rise of digital cash programs will provide a convenient mechanism for paying refugees for their labor.

Finally, while this volume has provided extensive discussion on data management and protection, with a focus on refugee rights in those processes, little attention has been paid to the serious challenges of cybersecurity and cyberwarfare. Certainly basic levels of security have been discussed (see chapter 7), however the deluge of data referred to repeatedly throughout this volume will bring greater challenges to the agencies collecting and managing those data. The more data they control, they greater the target they present. Questions such as the impact of cyberwarfare on forced migration, how cyberwarfare could be the new form of camp militarization, and how cross-border cyberwarfare will affect refugees are only now being considered. UN OCHA's (2014) policy paper titled "Humanitarianism in the Age of Cyber-Warfare" calls for establishing a "humanitarian cyberspace," protecting humanitarian organizations and their displaced partners from attacks. Future work in this domain can build upon a long history of scholarship in the relationship between refugee movements and state security. For an overview see Milner (2009, pp. 61–83).

Refugee Perspectives

While the research agendas defined in various chapters and synthesized above will generate benefits for refugees and may in fact involve refugees (see Fisher, chapter 5 and Tomaszewski, chapter 8), the primary focus has been on humanitarian organizations. This is due in part to their role in using technologies in refugee service provision as well as their role in providing technologies, such as SIM cards and mobile phones, directly to refugees.

Regardless, this agenda is very different from one focused exclusively on, for example, the significant role of refugee innovation in ICTs. There are many interesting developments in the ways refugees are helping themselves and one another through mobile technologies and social media. Indeed, an entire book could be dedicated to refugees' use of social media. There is fascinating work to be done on the hidden uses of ICTs, how refugees and the displaced use ICTs to cope, adjust, resist authorities—including humanitarian organizations—and in some cases thrive. And while a wide range of refugee-centric studies are needed, there is also a need for studies of how their use changes the global refugee-support system. As refugees continue to adopt and use ICTs in unique ways, their self-reliance and relationship with service providers are likely to change.

Implementing the Agenda

The above discussion highlights the breadth of research to be undertaken in the area of ICTs for refugees and displacement. In the following sections, tips for implementing the agenda put forth here, but also relevant for some of the areas discussed above in the limitations, are proposed. In particular, the discussion focuses on the dimensions and benefits of collaboration, as well as ethical issues and practicalities that may be encountered in the conduct of refugee research.

Collaboration

Conducting research on, for, and with refugees can be done independently, with appropriate permissions, or through collaboration with humanitarian organizations. In collaborative endeavors, organizations can help identify subjects, and, more importantly, provide insight on the circumstances, current and past, of a particular group of refugees. This is not to say humanitarian organizations understand the entire picture—nor are they free from bias. However, researchers who recognize potential biases, and find the role they are ascribed by working with a humanitarian agency does not conflict with their research goals, are likely to find collaboration helpful.

Whether collaborating or not, refugee research typically requires permissions for conducting research with a vulnerable population. This may require special permits above and beyond normal research permits, for example, from ministries or government agencies in charge of refugee affairs. If research is to be conducted in settlements or camps, special permission may also be required. As camp managers, national governments, or UNHCR in the case of refugees and IOM in the case of IDPs, are likely to be in charge of granting or coordinating such permits.

Conducting research on urban refugees can be easier for several reasons. First, permission to enter a camp, which may be constrained due to security concerns, is not necessary. Second, since camps tend to be located in remote areas, closer to international borders and in some cases armed conflicts, urban centers can simply be easier to reach. However, urban refugee research also has its challenges. One is their varied locations. Even the humanitarian organizations that know and are in frequent contact with refugees also find it difficult to pinpoint their homes and locations. It is more likely research will be conducted with those who visit humanitarian agency offices, but this will not provide a representative sample if one is required.

In addition, if a researcher is neither from the country from which the displaced are fleeing nor the host country, they may want to seek collaboration with academics from either country. The insights to be gained are critical and greatly enhance the research at each stage. Research on ICTs often involves the daily lives and experiences of the displaced. Researchers with knowledge of local conditions can help provide insight on the sensitivity and potential challenges questions create, as well as the likely validity of the answers.

Finally, sometimes the best insights can be derived from working with refugees themselves. It is often the case that refugee communities, encountered in any of the phases of the lifecycle, are extremely diverse, some with little formal education, yet others with advanced degrees, with each group having unique experiences and skills. As noted by Temple and Moran (2006) in the introduction to their edited volume titled *Doing Research with Refugees: Issues and Guidelines*, refugees can be engaged in a variety of types of research in a variety of ways. They encourage, as would the authors of this volume, participation in meaningful ways that value contributions across all phases of a research project, including fundamental aspects such as framing and defining research questions, as well as analyzing and critiquing results. Engagement must, unfortunately, take into account the challenges faced by many refugees in terms of family matters and time needed to deal with authorities. Hence, researchers must be sensitive in their engagement.

Finally, researchers must be cognizant of the need to manage expectations. Many humanitarian organizations, let alone refugees themselves, are unaware of the nature of academic research, which typically seeks to develop and extend theories or prototype devices in testing new designs. Further, academic research typically takes longer than an aid or development program. With different goals and timeframes, it is rare for academic research to solve problems in the short term. Hence, it is critical that the ultimate goals of the research and its timeline be clearly articulated. Researchers working with humanitarian organizations may find it necessary to compromise, integrating short-term goals, and establishing preliminary outcomes and reports in order to meet the needs of collaborators.

Ethics and Practicalities

Conducting research on any vulnerable population must be approached with sensitivity and an eye toward a variety of ethical issues. As when dealing with any community that has faced persecution, even an issue as basic as the purpose and conduct of the research needs to be carefully considered. In some cases, pursuit of fundamental scientific advances through refugee research may be problematic. For example, testing novel technologies or theories that neglect the impact on or interaction with refugees' traumatic experiences should be avoided. However, fundamental or basic research that is carefully designed has the potential to generate both helpful information and technical solutions, while at the same time providing the evidence for fundamental scientific advances. Oftentimes, consultation with research protections professionals in universities or with humanitarian organizations tasked with protection can be helpful in finding a middle ground.

In some cases, the middle ground might require compromise on a scientific question or taking on extra work. For example, deploying novel network technologies where no communication networks are available can be problematic in that subjects, in this case users, experience harm from an unreliable technology. However, if existing technology can be deployed at the same time, creating an experimental design comparing new and established technology use, at least some subjects benefit right away, and if the established equipment is donated, all subjects eventually benefit.

Ethical concerns must be contemplated not only in the design of the research but also in the data collection necessary for some social scientific or technical research. Managing expectations is just one element of an ethical approach. For example, research that engages directly with refugees in testing new technologies must be clear as to whether or not refugees will have ongoing access to prototypes, as well as provide realistic information as to the likelihood of the technology ever becoming commercially available.

Similarly, researchers merely interacting with refugees for data collection must also be guarded about their subjects' expectations. In this realm, refugee researchers can learn from the ICTD community. As Wyche et al. (2015) note in their work engaging with mobile phone repair workers in Kenya, the repair technicians saw Wyche, a white woman from the U.S., as likely to have direct connections with companies such as Nokia or Samsung. This expectation may have arisen from a research design that had the technicians draw designs for mobile phones they felt would be more robust. Naturally, the technicians wanted their concepts delivered to these firms. The challenge for researchers is to anticipate these expectations and explain the likely outcomes very clearly.

Another important lesson from the ICTD and social science communities in general, is to be aware of the role research participation plays in the lives of subjects, particularly where payment is offered. The issue was brought to light again by a recent story in the U.S. popular press on the "research and/or development economy" of the Kibera settlement in Kenya.[4] For people with few employment opportunities, or no legal employment opportunities as may be the case for refugees, participating in research or development programs may be their primary and perhaps only employment. If honest responses are provided, *in and of itself* this should not pose a problem. However, researchers need to be aware and assess the consequences for their study.

General guidance on ethics can be gleaned from writings on the ethics of ICTD or ICT4D research (see e.g., Sterling & Rangaswamy, 2010; and Dearden, 2013). Issues such as power differentials, mutual respect, and the need to thoroughly investigate context are common across development and refugee research. Furthermore, similar to ICTD researchers, refugee researchers must be careful during their engagements to establish and implement practices of professional conduct. The rise of "poverty tourism" can call into question the motives of preliminary field visits or exploratory research. Despite these commonalities, the special circumstances of refugees call for increased attention to ethics. In particular, refugees' and IDPs' experiences of persecution and general lack of rights call for increased attention to issues of exploitation and potential harms (Hugman & Bartolomei, 2014; Leaning, 2001; Nunn, McMichael, Gifford, & Correa-Velez, 2014).

Finally, adequate protections for subjects requires an understanding of the broader context of the refugee subject population, including the conditions from which they fled and those where they currently reside, as the entire history will inform an assessment of the extent of harms. It may be the case that university officials do not have the information readily available or experience necessary to quickly assess a research protocol. To smooth the process, researchers are advised to engage research protections staff early in order to help them identify qualified reviewers. If the research is being carried out with humanitarian or other knowledgeable organizations or individuals, university research protections boards may be willing to accept their judgments or inputs to the process as well.

Conclusion

The confluence of forced migration and ICT use is poised to change the experience of displacement as well as the sector organized to address the resulting trauma, poverty and hopelessness. As a common driver, mobility underpins both the technical architecture and the response to crises ranging from climate change to armed conflict. The result is a greater need for information, which is being supported by a plethora of data-related innovations.

The changes these trends are propelling, envisioned in these chapters, call for a dedicated research agenda. When conducted with foresight, as enabled by this volume, research can shape the direction of technological change as well as its impacts.

The research agendas (proposed here and in each of this volume's chapters) offer many insights to displacement scholars, practitioners, and advocates alike. For scholars, the agenda not only identifies but also justifies interesting lines of inquiry, in addition to situating them in the relevant scholarship. The agenda also presents opportunities for interdisciplinary approaches, which have become popular in some universities and with some funding agencies. Relatedly, funding agencies working with scholars and practitioners may find the research agenda helpful in identifying new opportunities for their existing programs. Foundations may also use the agenda to develop a comprehensive approach to ICTs for displacement. We hope this volume might also spur cross-agency programs, combining the benefits of fundamental science with actionable results obtained from more applied research.

For practitioners, the book spans various stages of the refugee lifecycle, allowing those working in temporary asylum to gain insight into the issues and roles of technology in, for example, refugee status determination (RSD). They will also gain insight into the uses of technology across a range of service areas, including RSD, registration, life in a camp, camp management, and food provision. The volume also provides practitioners insights into trends, such as the emergence of the digital refugee and digital humanitarian brokerage. These insights can help envision new programs and chart new career paths. For practitioners engaged in developing technologies for refugees or humanitarian organizations, this volume provides insights into mechanisms of adoption and scaling critical for new technology success.

Finally, as opposed to a contemporary treatment of ICTs for displacement, this volume's future orientation provides two important benefits. First, as both forced migration and technological change will continue for the foreseeable future, the research topics proposed herein can help steer technological developments toward trajectories that benefit refugees. Second, as some of the most vulnerable people on earth, the displaced need independent voices who can analyze and point to areas in which their rights are likely to be ignored or infringed upon. Forward looking analyses anticipating the effects of new technologies can provide an early start for defining new policies to ensure their protection. As such, these efforts contribute to ongoing analyses of the effects of ICTs for refugees and the displaced, testing their strengths and exposing their weaknesses as a digital lifeline.

Notes

1. https://www.fastcompany.com/3016393/how-the-un-uses-crowdsourcing-to-get-refugees-what-they-need.

2. https://challenges.openideo.com/challenge/refugee-education/research/the-world-s-first-digital-fabrication-lab-fab-lab-in-a-refugee-camp.

3. https://www.cnet.com/news/bringing-tech-jobs-to-third-world-refugees/

4. http://www.marketplace.org/2016/07/28/world/ngos-nairobi-have-pay-locals-attend-meetings.

References

Alam, K., & Imran, S. (2015). The Digital Divide and Social Inclusion Among Refugee Migrants. *Information Technology & People, 28*(2), 344–365. Retrieved from doi:10.1108/ITP-04-2014-0083.

Andrade, A. D., & Doolin, B. (2016). Information and Communication Technology and the Social Inclusion of Refugees. *Management Information Systems Quarterly, 40*(2), 405–416.

Conneally, P. (2011). *Digital Humanitarianism.* TED Talks Usage Policy. Retrieved from https://www.youtube.com/watch?v=L9_c1j9VRwE

Dearden, A. (2013). See No Evil: Ethics in an Interventionist ICTD. *Information Technologies and International Development, 9*(2), 1–17.

Doná, G. (2015). Making Homes in Limbo: Embodied Virtual "Homes" in Prolonged Conditions of Displacement. *Refuge: Canada's Journal on Refugees, 31*(1), 67–74.

Hellmann, D. E., Maitland, C. F., & Tapia, A. H. (2016). Collaborative Analytics and Brokering in Digital Humanitarian Response. In *Proceedings of the ACM 2016 Conference on Computer Supported Cooperative Work—CSCW '16* (pp. 1284–1294). doi:10.1145/2818048.2820067.

Hugman, R., & Bartolomei, L. (2014). The Ethics of Participation in Community Work Practice. In A. K. Larsen, V. Sewpaul, & G. O. Hole (Eds.), *Participation in Community Work: International Perspectives* (pp. 19–29). New York: Routledge.

Jacobsen, K. L. (2005). *The Economic Life of Refugees*. Boulder, CO: Kumarian Press.

Jacobsen, K. L. (2015). Experimentation in Humanitarian Locations: UNHCR and Biometric Registration of Afghan Refugees. *Security Dialogue, 46*(2), 144–164. Retrieved from doi:10.1177/0967010614552545.

Leaning, J. (2001). Ethics of Research in Refugee Populations. *Lancet, 357*(9266), 1432–1433.

Meier, P. (2015). *Digital Humanitarians: How Big Data Is Changing the Face of Humanitarian Response*. Boca Raton, FL: CRC Press.

Milner, J. (2009). *Refugees, the State and the Politics of Asylum in Africa*. New York: Springer.

Moorehead, C. (2006). *Human Cargo: A Journey Among Refugees*. London: Vintage.

Nunn, C., McMichael, C., Gifford, S. M., & Correa-Velez, I. (2014). "I came to this country for a better life": Factors mediating employment trajectories among young people who migrated to Australia as refugees during adolescence. *Journal of Youth Studies, 17*(9), 1205–1220.

Robertson, Z., Wilding, R., & Gifford, S. (2016). Mediating the Family Imaginary: Young People Negotiating Absence in Transnational Refugee Families. *Global Networks, 16*(2), 219–236.

Sterling, S. R., & Rangaswamy, N. (2010). Constructing Informed Consent in ICT4D. In proceedings of the *4th ACM/IEEE International Conference on Information and Communication Technologies and Development* (pp. 1–9). ACM (Association for Computing Machinery).

Temple, B., & Moran, R. (2006). *Doing Research with Refugees: Issues and Guidelines*. Bristol, UK: Policy Press.

UN OCHA (Office for the Coordination of Humanitarian Affairs). (2013).*Humanitarianism in the Network Age*. UN OCHA Policy and Studies Series.

UN OCHA (Office for the Coordination of Humanitarian Affairs). (2014, October 20). Humanitarianism in the Age of Cyber-warfare: Towards the Principled and Humanitarian Emergencies. *OCHA Policy and Studies Series*. Retrieved from http://reliefweb.int/sites/reliefweb.int/files/resources/Humanitarianism in the Cyberwarfare Age—OCHA Policy Paper 11.pdf.

Whittaker, D. J. (2006). *Asylum Seekers and Refugees in the Contemporary World*. New York: Routledge.

Wilding, R., & Gifford, S. M. (2013). Special Issue: Forced Displacement, Refugees and ICTs: Transformations of Place, Power and Social Ties Introduction. *Journal of Refugee Studies, 26*(4), 495–504.

Witteborn, S. (2015). Becoming (Im)Perceptible: Forced Migrants and Virtual Practice. *Journal of Refugee Studies, 28*(3), 350–367. Retrieved from doi:10.1093/jrs/feu036.

Wyche, S., Dillahunt, T. R., Simiyu, N., & Alaka, S. (2015). "If God Gives Me The Chance I Will Design my Own Phone": Exploring Mobile Phone Repair and Postcolonial Approaches to Design in Rural Kenya. In proceedings of the *2015 ACM International Joint Conference on Pervasive and Ubiquitous Computing*. ACM (Association for Computing Machinery).

Contributors

Elizabeth Belding

Elizabeth Belding is a Professor in the Department of Computer Science at the University of California, Santa Barbara (UCSB). Her research focuses on mobile networking, including network performance analysis and information and communication technology for development (ICTD). She is the author of over 100 technical papers and has served on over 60 program committees for networking conferences. She is currently an Editor-at-Large for IEEE Transactions on Networking. She is an IEEE Fellow and an ACM Distinguished Scientist.

Karen E. Fisher

Karen E. Fisher is Professor, Information School; Adjunct Professor, Communication Department, University of Washington, Seattle. She is Consultant for UNHCR Jordan; Visiting Professor at the Open Lab, Newcastle University, UK; and Adjunct Professor, Åbo Akademi University, Turku, Finland. An advocate of humanitarian research, her passion is how Human Computer Information (HCI)-industry-NGO collaborations can improve lives around the world and build futures through participatory methods. Her priority is working with youth and families to increase educational opportunity, livelihoods, and social connectedness at the UNHCR Za'atari Refugee Camp in Jordan and other Middle East settings. Parallel fieldwork focuses on refugees in the European Union, understanding their information worlds, and the economic and social/political impacts of migration. With a mantra of "Youth First," Dr. Fisher's InfoMe group conducts in situ, co-design workshops with teens to understand how youth feel about social justice, and how they help families, friends, and institutions as ICT wayfarers, and are redefining information and the nature of information work through visual media. Dr. Fisher is renowned for her development and use of theory and methods for understanding information problems, specifically how people experience information via everyday life, with interpersonal aspects and role of informal social settings or "Information Grounds." With colleagues, Dr. Fisher spearheaded several landmark projects, including the U.S. Impact Study for the Bill & Melinda Gates Foundation of how people use technology in public libraries. Her *Theories of Information Behavior* (2005) remains the top-selling monograph at http://www.asist.org, with numerous papers in most cited lists about how people engage with information. With a doctoral degree in Information Science from the University of Western Ontario and a postdoctorate from at the University of Michigan, Dr. Fisher's work is supported by UNHCR, Google, National Science Foundation, Amazon, LEGO Foundation, Microsoft, Institute of Museum & Library Services, the Bill & Melinda Gates Foundation, and the Social Sciences and Humanities Research Council of Canada. Website: http://syria.ischool.uw.edu.

Daniel Iland

Daniel Iland is a Senior Software Engineer on Sensing, Inference, and Research at Uber. His work primarily focuses on using sensor fusion to improve mobile device location accuracy. At Uber, he helped form a cross-organizational team to coordinate Uber's technology response to natural disasters. As a PhD student in the Department of Computer Science at the University of California, Santa Barbara (UCSB), his research focused on enabling robust and reliable communication in emergency and disaster scenarios, wireless localization, and ICTD. He was awarded a bachelor's degree in Computer Science from the Rochester Institute of Technology in 2011.

Lindsey N. Kingston

Lindsey Kingston is an Associate Professor of International Human Rights at Webster University in Saint Louis, Missouri. She directs the university's Institute for Human Rights and Humanitarian Studies and is an expert on the issues of statelessness and forced displacement. Her research has taken her to fieldwork locations around the world—including Rwanda, the Canadian Arctic territory of Nunavut, Eastern Europe, and the Mediterranean region. Kingston is an editor for *Human Rights Review*, and her work has been published in a variety of edited volumes and peer-reviewed journals, including *Journal of Human Rights, Forced Migration Review, Journal of Human Rights Practice, Peace Review,* and *The International Journal of Human Rights.* She earned her PhD in Social Science from Syracuse University's Maxwell School of Citizenship and Public Affairs in 2010; her doctoral dissertation on statelessness and issue emergence earned special recognition from the United Nations High Commissioner for Refugees.

Carleen F. Maitland

Carleen Maitland is an Associate Professor in the College of Information Sciences and Technology at Pennsylvania State University, where she also serves as Co-director of the Institute for Information Policy. She is an expert in humanitarian informatics. Her work has been carried out in the U.S., Europe, Africa, and the Middle East, while working with diverse organizations such as the UN Conference on Trade and Development (UNCTAD), the UN Office for Coordination of Humanitarian Affairs (UN OCHA), the UN Refugee Agency (UNHCR), Save the Children, and the U.S. State Department, to name a few. Her work has been supported by the U.S. National Science Foundation (NSF), the U.S. Department of Commerce, the European Commission, and the United Nations, among others. From 2010 to 2012, she served as Program Officer in NSF's Office of International Science and Engineering and its Office of Cyberinfrastructure. She holds BS and MS degrees in engineering, from Worcester Polytechnic Institute and Stanford University, respectively, and a doctorate in the Economics of Infrastructure from the Technical University of Delft in the Netherlands.

Susan F. Martin

Susan Martin is the Donald G. Herzberg Professor Emeritus in the School of Foreign Service at Georgetown University. She previously served as the Director of Georgetown's Institute for the Study of International Migration. She currently serves as the Chair of the Thematic Working Group on Environmental Change and Migration for the Knowledge Partnership in

Migration and Development (KNOMAD) at the World Bank. Before coming to Georgetown, Dr. Martin was the Executive Director of the U.S. Commission on Immigration Reform and Director of Research and Policy at the Refugee Policy Group. Dr. Martin received her MA and PhD from the University of Pennsylvania and her BA from Douglass College, Rutgers University. She previously taught at Brandeis University and the University of Pennsylvania.

Galya Ben-Arieh Ruffer

Galya Ben-Arieh Ruffer, JD, PhD, is the founding Director of the Center for Forced Migration Studies at the Buffett Institute for Global Studies, Northwestern University. Her research centers on the rights and processes of refugee protection and inclusion in host societies. She has received private funding to launch a research program on refugee resettlement, has been awarded grants from the National Science Foundation, the Social Science Research Council, and the Kellogg Center for Dispute Resolution, and is a former Senior Fellow at the Käte Hamburger Kolleg/Centre for Global Cooperation Research (University of Duisburg-Essen). She has conducted field research in the Great Lakes region of East Africa, Germany, and the U.S., and has published on testimony and justice, asylum law and policy, refugee protection in a digital age, human rights litigation in transnational courts, and immigrant incorporation and integration in Europe, with a recent book, *Adjudicating Refugee and Asylum Status: The Role of Witness, Expertise, and Testimony* (co-edited with Benjamin Lawrance), Cambridge University Press (2015). She is part of the Forced Migration Upward Mobility Project team to rethink refugees in resettlement as active agents in their own livelihoods, serves on the executive committee of the International Association for the Study of Forced Migration, and has worked as an immigration attorney representing political asylum claimants both as a solo-practitioner and as a pro-bono attorney. She completed a JD at Northwestern University and a PhD in Political Science at the University of Pennsylvania.

Paul Schmitt

Paul Schmitt is a Postdoctoral Research Associate at the Center for Information Technology Policy at Princeton University. He received his PhD in

Computer Science from the University of California, Santa Barbara (UCSB). His research focus is on network systems design, and network measurement and performance analysis. His work spans a wide range of topics, including local cellular networks, commercial cellular data core characterization, wireless spectrum sensing, and network connectivity in resource-limited environments.

Lisa Singh

Lisa Singh is a Professor in the Department of Computer Science and a Research Affiliate of the Mass Data Institute at Georgetown University. She has authored/co-authored more than 50 peer reviewed publications and book chapters. Her research crosses different aspects of data-centric computing, including data science, data mining, and data privacy. Her research has been supported by different government agencies, including the National Science Foundation, the Office of Naval Research, the Social Science and Humanities Research Council, and the U.S. Department of Defense. Dr. Singh has also been involved with organizing multiple workshops involving future directions of big data research. She is also currently involved in different organizations working on increasing participation of women in computing and integrating computational thinking education into K–12 curricula. She received her BSE degree from Duke University and her MS and PhD degrees from Northwestern University.

Brian Tomaszewski

Brian Tomaszewski, PhD, is a Geographic Information Scientist with research interests in the domains of Geographic Information Science and Technology, Geographic Visualization, Spatial Thinking, and Disaster Management and Refugee Affairs. His published research on Geographic Information Systems (GIS) and Disaster Management-related topics has appeared in top scientific journals and at conferences, such as *Information Visualization, Computers, Environment and Urban Systems, Computers and Geosciences*, the *IEEE Conference on Visual Analytics Science and Technology*, the *IEEE Global Humanitarian Technology Conference* and *The Cartographic Journal*. He is the author of *Geographic Information Systems for Disaster Management*, a leading textbook on the topic, published in 2015 by Taylor and Francis. His relevant experience includes past work with internationally focused organizations interested in GIS and disaster management, and refugee issues, such as the United Nations Office for the Coordination for Humanitarian Affairs (UN OCHA) ReliefWeb service, United Nations Office for Outer Space Affairs Platform for Space-based Information for Disaster Management and Emergency Response (UN-SPIDER), United Nations Global Pulse, United Nations High Commissioner for Refugees (UNHCR), and the United Nations Institute for Environment and Human Security (UNU-EHS). His research has been funded by the U.S. National Science Foundation (NSF), United Kingdom Department for International Development (UK-DFID), and Grand Challenges Canada. He is an internationally focused scholar with active disaster management and refugee research activities in Germany, Rwanda, Jordan, and India. Dr. Tomaszewski is currently an Adjunct Professor with the Centre for Disaster Management and Mitigation, Vellore Institute of Technology, India, and an Associate Professor in the Department of Information Sciences & Technologies at the Rochester Institute of Technology. He holds a PhD in Geography from Pennsylvania State University. For more information, visit: https://www.rit.edu/gccis/geoinfosciencecenter.

Mariya Zheleva

Mariya Zheleva received her PhD in Computer Science from the University of California, Santa Barbara (UCSB) in 2014. She is an assistant professor in the Department of Computer Science at University at Albany SUNY. Dr. Zheleva's research interest is in the intersection of wireless networks and Information and Communication Technology for Development. She has done work on small local cellular networks, Dynamic Spectrum Access, spectrum management and sensing, and network performance and characterization. She is the founder and director of the UbiNet Lab at University at Albany.

Index

Note: page numbers followed by "f," "t," and "n" refer to figures, tables, and endnotes, respectively.

Printed in the United States
by Baker & Taylor Publisher Services